D1168386

Venus

Venus

Patrick Moore

First published in Great Britain in 2002 by Cassell Illustrated,
a division of Octopus Publishing Group Limited,
2-4 Heron Quays, London E14 4JP

Copyright © Patrick Moore 2002.

The right of Patrick Moore to be identified
as the author of this work has been asserted
by him in accordance with the Copyright,
Designs and Patents Act of 1988.

All rights reserved. No part of this publication may be
reproduced, stored in a retrieval system, or transmitted in any
form or by any means, electronic, mechanical, photocopying,
recording, or otherwise, without the prior permission of the
publisher.

A CIP catalogue record for this book is available
from the British Library

ISBN 0 304 36281 6

Printed in Slovenia by DELO tiskarna
by arrangement with Prešernova družba

Gula Mons on Venus – Magellan radar image (courtesy JPL).

Contents

Introduction

Ask the average person which is the most interesting of all the planets, and the reply is almost certain to be 'Mars'. The reason is that of all the planets in the Solar System, Mars is the least unlike the Earth. True, it is smaller and colder, but it does have an appreciable atmosphere, and we can see the icy polar caps, together with the reddish 'deserts' and the dark areas once believed to be old sea-beds filled with vegetation. Only a few decades ago it was widely believed that an advanced civilization flourished there, and that the Martians had built an elaborate canal system to transport water from the polar zones through to the centres of population closer to the equator.

The Martians have been banished to the realm of myth, but there may still be primitive life forms, and plans are already being made to send manned spacecraft there. But what about Venus, our sister world? Larger than Mars, closer to the Sun than we are, and surrounded by a dense, cloud-laden atmosphere which hides the actual surface completely. Here we have a very different situation.

Before the Space Age, we knew very little about conditions on Venus, and we did not even know the length of the planet's "day". For all we knew, conditions there might be quite tolerable, with broad oceans and luxuriant vegetation. It was suggested that as a potential colony, Venus might well be more promising than Mars. In my book *The Planet Venus*, published in 1954, I wrote: "It is dangerous to state that Venus is bound to be totally lifeless. Some 500 million years ago, in Cambrian times, the Earth was a world containing vast stretches of ocean, and in the waters flourished

NASA's Saturn Explorer Cassini, with ESA's Titan Probe Huygens attached, being successfully launched at Cape Canaveral 16 October 1997.

primitive organisms which were later to evolve into land creatures, mammals and finally man. Vegetation on the lands had not appeared, and even insects lay in the future. Conditions of this sort fit in quite well with the 'marine' theory of Venus. ... It is possible, therefore, that Venus is a world in a 'Cambrian' state, and that the primitive marine organisms, if they exist, will eventually develop into more advanced forms of life."

When the rockets began to fly, the Russians took the lead, and, significantly, they concentrated upon Venus. American interest was equally keen, but attitudes changed when it was found that Venus is overwhelmingly hostile. Images sent back direct from the surface, below those dense, fuming clouds, showed that manned expeditions were out of the question, at least in the foreseeable future. Not surprisingly, the space-planners turned back toward Mars, and Venus was somewhat neglected. Between 1961 and 1983, twenty-three probes were launched with Venus as the main target. Since then, there has been only one.

Yet Venus is a fascinating world, and we can learn a great deal from it. In this book – very different from its forerunner of 1954 – I have tried to present a balanced picture, dealing with basic facts, history, current research, and what we may expect in the years to come. There is every chance that new spacecraft will be sent there before long, in which case Venus will once more become a focus of attention. And even non-astronomers cannot ignore it; after all, apart from the Sun and Moon, it is the most brilliant object in the entire sky.

1

Non-Identical Twin

One evening in early January 2001 – the start of the new millennium – I went out to my observatory. The Sun had only just set; the sky was darkening, and there, in the south-west, was the planet Venus. It was at its very best, looking like a small lamp; later, when night had fallen, it would be capable of casting perceptible shadows. It was much the brightest object in the sky apart from the Sun and the Moon, and it is small wonder that the ancients named it in honour of the Goddess of Beauty.

Yet we have to admit that appearances are deceptive. Far from being a friendly, welcoming world, Venus is a veritable inferno with a scorching hot surface, a thick, choking atmosphere, and clouds rich in sulphuric acid. In size and mass it is almost a twin of the Earth, but it is very definitely a non-identical twin. In most respects they are wildly dissimilar.

The main reason for this is that Venus is much closer to the Sun than we are. The Earth is 93,000,000 miles from the Sun, Venus only 67,000,000 miles; the difference may be less than 30,000,000 miles, but it is all-important. The story goes back to the very beginning of what we now call the Solar System.

The Sun is a star; a globe of incandescent gas, vast enough to contain well over a million bodies the volume of the Earth, and shining because of nuclear reactions taking place inside it. The most plentiful substance in the universe is hydrogen, and the Sun contains a great deal of it. Near the centre of the globe, where the temperature is of the order of 15,000,000 °C (27 million °F) and the pressures are colossal, the nuclei of hydrogen atoms are

The Moon and Venus, photograph by H.J.P. Arnold.

combining to build up nuclei of the second most plentiful element, helium. It takes four hydrogen nuclei to make up one nucleus of helium, and every time this happens a little energy is set free and a little mass – or "weight", if you like – is lost. It is this energy which makes the Sun shine, and the loss in mass amounts to 4,000,000 tons per second. This may sound staggering, but by solar standards it is trivial. The Sun was born about 5,000 million years ago, and it will not change much for a long time yet, though of course it will not last for ever. For the moment it is the absolute ruler of the Solar System, and we are completely dependent on it.

The Sun condensed out of a cloud of dust and gas in space, and at first it was surrounded by a "solar nebula", also made up of dust and (mainly) gas. The planets grew up by accretion inside this solar nebula, and at least we are confident about the timescale, because we know that the age of the Earth is about 4,600 million years.* Venus is certainly about the same age, and since it is our twin it is logical to assume that the two worlds began to evolve along similar lines – perhaps with similar atmospheres, similar

Venus against the Hyades, the bright star cluster in the constellation Taurus.

Painting by Paul Doherty.

Crescent Venus, photographed against a dark sky from Long Crendon Observatory, Oxfordshire, in early 2001 (Gordon Rogers).

lands and similar seas. But one factor then caused a major change. Originally, the Sun was not nearly so luminous as it is now. When its power increased, the Earth was out of harm's way; Venus, much closer-in, was not. Therefore the oceans of Venus boiled away, the carbonates were driven out of the rocks, and the atmosphere became loaded with carbon dioxide. This gas shuts in the Sun's heat, and leads to what we may well call a greenhouse effect, so that in a relatively brief time, cosmically speaking, Venus changed from a potentially life-bearing world into the furnace-like inferno of today. Any life which may have appeared there was ruthlessly snuffed out. I will have more to say about this later; for the moment it is enough to say that advanced life of the kind we know cannot possibly exist there. The Goddess of Beauty is unpleasantly hostile.

This does not make Venus any the less interesting, and it is of course one of the worlds close enough to be reached by our present-day rockets. But first it will, I think, be helpful to say a little about the Solar System itself, and to see where Venus fits into the general pattern.

---

* This is often given as 4.6 billion, but I always avoid the term "billion" because there is a slight danger of confusion. The American billion, now generally accepted, is equivalent to one thousand million, while the old Imperial billion is equal to one million million.

2

Venus in the Solar System

The Solar System is our home in space, and we naturally regard it as important. In fact, it is important only to us. The Sun is a very ordinary star, one of 100,000 million suns in our own particular star-system or Galaxy – and there are thousands of millions of galaxies, so that the total number of stars in the universe is beyond our comprehension. (There are many more stars than there are grains of sand in the Sahara Desert.) We are now certain that many of these stars have planetary families, so that there is nothing unusual about the Solar System – except that we happen to live in it.

The planets have no light of their own; they shine only by reflecting the rays of the Sun. Outwardly they look like stars, but unlike the stars they do not keep to the same relative positions. The stars are so far away that their individual or "proper" motions are very slight, and the patterns or constellations we see now are to all intents and purposes the same as those which must have been seen by King Canute, Julius Caesar and the builders of the pyramids; even the nearest star beyond the Sun is over 24 million million miles away. Relatively speaking, the planets are our near neighbours, and they shift very obviously against the starry background. This was how they were first identified, and five of them (Mercury, Venus, Mars, Jupiter and Saturn) are bright naked-eye objects, though the outermost members of the family (Uranus, Neptune and Pluto) have been discovered only in near-modern times. The names, of course, are mythological, and were originally Greek, though we use the Latin equivalents. Thus the war-god,

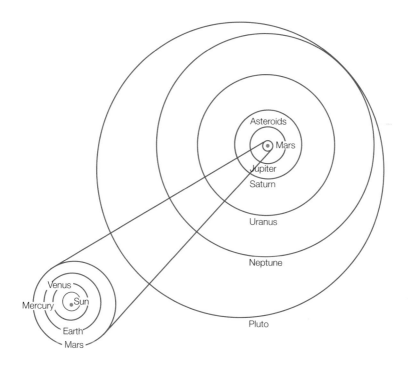

Plan of the Solar System – orbits of the inner planets have been pulled out to show them clearly because they are so small compared to those of the outer worlds.

known to the Greeks as Ares, is our Mars, while Zeus, the ruler of Olympus, has become Jupiter, and the Greek Aphrodite is our Venus.

En passant, there is no really good adjective for Venus. I have always liked "Cytherean", from an old Sicilian name for the goddess, but "Venusian" is now generally used. It is ugly, but "Venerean" is even worse. Therefore, I propose to use "Venus" both as a noun and as an adjective.

Here are the main data for the planets:

Name	Mean distance from Sun, (million miles)	Orbital Period	Rotation Period	Diameter (miles)	Mass (Earth=1)	Escape Velocity miles/s
Mercury	36	88 days	58.6 days	3,030	0.055	2.6
Venus	67	224.7 days	243.02 days	7,523	0.815	6.4
Earth	93	365.2 days	23h 56m	7,926	1	7.0
Mars	142	687 days	24h 37m	4,222	0.107	3.2
Jupiter	483	11.9 years	9h 50m	89,424	318	37
Saturn	886	29.5 years	10h 14m	74,914	95	22
Uranus	1,783	84.0 years	17h 14m	31,770	15	14
Neptune	2,793	164.8 years	16h 7m	31,410	17	15
Pluto	3,666	247.7 years	6d 9h	1,444	0.002	0.7

No, there is no misprint in the table! Venus' rotation period or "day" really is longer than its "year"; I will have more to say about this later.

Most of the planets have satellites moving round them. We have one – our familiar Moon, whose diameter is 2,160 miles, and which is illuminated by reflected sunlight. Of the rest, Mars has two dwarf satellites; Jupiter has a true family, of which four members are large; Saturn also has an extensive family, though only one member of it (Titan) is as large as our Moon; Uranus has over twenty satellites, Neptune eight and Pluto one. Only Mercury and Venus are solitary travellers in space.

A casual glance at a plan of the Solar System is enough to show that it is divided into two sharply defined parts. We start with four relatively small, solid worlds: Mercury to Mars. Beyond we come to a belt of midgets known as minor planets or asteroids, only one of which (Ceres) is as much as 500 miles in diameter, and only one of which (Vesta) is ever visible with the naked eye. There follow the four giants (Jupiter to Neptune), which have gaseous surfaces and relatively small, solid cores. Finally there is Pluto, which is very much of a maverick, and may not be worthy of true planetary status. Many other dwarfs, of asteroidal size, move round the Sun

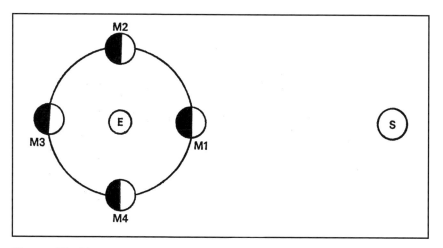

Phases of the Moon.

at distances comparable with that of Pluto. They have been found only during the last decade; they are so far away that they are very faint.

It is fitting to begin our lightning tour of the Sun's family with the Moon, which is our faithful companion, and moves at a mean distance of less than a quarter of a million miles from us – 239,000 miles on average, with an orbital period of 27.3 days. Because it shines only by reflected sunlight, it shows regular phases, or changes of shape, from new to full. The diagram above should make this clear. In position 1, the Moon is more or less between the Earth and the Sun; its unlit side is turned toward us at "new moon", and we cannot see it unless the lining-up is exact, when the Moon will pass directly in front of the Sun and produce a solar eclipse. By sheer chance, the Sun and the Moon appear virtually the same size in the sky. The Sun's diameter is 400 times that of the Moon, but the Sun is also 400 times further away. If the brilliant solar disk is completely covered, at a total eclipse, the spectacle is glorious. The Sun's atmosphere – the corona – flashes into view, perhaps with what are termed prominences, masses of glowing gas rising from the Sun's surface. The sky darkens, and planets and bright stars can be seen; Venus, if suitably placed, is outstanding. Unfortunately the Moon's shadow is only just long enough to touch the Earth, so that to see a total solar eclipse you have to be in just the right place at just the right time. From England, the last chance was on 11 August 1999, when the track of totality crossed the West Country. (I was at Falmouth in Cornwall, sheltering under an umbrella and saying unkind things about the clouds and rain.) Anyone staying in England must wait until 23 September 2090, though other parts of the world are more favoured.

The Moon is a world of mountains, valleys, great plains, miscalled "seas", and above all craters, ranging from tiny pits up to vast enclosures well over 150 miles in diameter. There is no water, and there is virtually no atmosphere. The Moon has a weak gravitational pull – its mass is only $^1/_{81}$ that of the Earth and any air it may once have had has long since leaked away into space. No life has ever appeared on the Moon; it has been sterile throughout its long history.

The Moon takes 27.3 days to complete one orbit. It spins on its axis in exactly the same time, so that it keeps the same face turned permanently toward us, and before 1959, when the Russians sent an unmanned space-craft on a "round trip", we had no direct information about the far side of the Moon. In fact, it proved to be just as rugged and just as barren as the side we have always known.

There is no mystery about this apparent coincidence. The Moon was formed at the same time as the Earth, 4.6 thousand million years ago. Either it and the Earth grew up in the same part of the original solar nebula, or else (as many astronomers now believe) the two bodies were initially combined, and the Moon produced as a result of a tremendous impact from an object which could have been at least a thousand miles across. At first the Moon rotated quickly, but its spin was slowed down by the gravitational pull of the Earth, which tried to keep a "bulge" in the still-viscous Moon turned in our direction. There is some analogy with a cycle-wheel which is rotating and is being steadily slowed by brake-shoes. Eventually the Moon's rotation, relative to the Earth, had stopped altogether; this is known as captured or synchronous rotation. Note, however, that the rotation relative to the Sun had not stopped, and day and night conditions are the same for both hemispheres – though from the far side, of course, the Earth cannot be seen, because it remains below the horizon.

Come now to the planets, which make up a varied family. Mercury and Venus, which are closer to the Sun than we are, are termed the "inferior planets", and show phases like those of the Moon. Mercury is not a great deal larger than the Moon, and has almost no atmosphere. It always keeps in the same part of the sky as the Sun, and so is never very conspicuous; at its best it may be seen with the naked eye either very low in the west after sunset or else very low in the east before dawn. It is never visible against a really dark background, and there are many people who have never seen it at all. Telescopes show very little on its surface, and our detailed knowledge comes from one unmanned spacecraft, Mariner 10, which flew past Mercury in 1974 and 1975, sending back excellent pictures. There are craters, mountains and valleys; the surface temperature can rise to over 375°C (700°F). The Mercurian "day" is long; the planet's rotation period is over 58 Earth-days, which is equal to two-thirds of Mercury's "year" of 88 Earth-days. The calendar there is most peculiar, but there are no local inhabitants to appreciate it; life on Mercury seems to be out of the question.

Beyond the orbits of Venus and Earth, those dissimilar twins, we come to the red world Mars, which can become really brilliant. Its colour is very pronounced, which is why it was named after the mythological God of War. In size it is intermediate between Mercury and the Earth; its "year" amounts to 687 Earth-days, equivalent to 668 Martian days or "sols", because the rotation period of Mars is just over half an hour longer than ours. It has a thin but appreciable carbon-dioxide atmosphere, with clouds and dust-storms. During summer on the Martian equator the temperature can rise to over 40 degrees Fahrenheit, but the atmosphere is very inefficient at retaining warmth, and the nights are bitterly cold.

Telescopes show a wealth of surface detail. When Mars is well placed – as, for example, in the summer of 2003 – even a modest telescope will show the red "deserts", the permanent dark

Venus photographed with the Hale 200-inch reflector on Palomar. For years the 200-inch was by far the most powerful telescope in the world. It must, however, be admitted that this picture of Venus is remarkably uninformative!

markings, and the white polar caps. The caps wax and wane with the Martian seasons, and are icy in nature, though they are made up of a combination of carbon dioxide ice and ordinary water ice. The dark markings are very prominent, and have been given names. The most conspicuous feature is the roughly triangular Syrtis Major, while in the northern hemisphere we have the almost equally prominent Acidalia Planitia. (The names were given in 1877 by the Italian astronomer Giovanni Schiaparelli, though they have been somewhat modified in recent times.) The so-called deserts are made up not of sand, but of reddish minerals; Mars is an arid, rusty world. The dark areas, once thought to be old sea-beds filled with vegetation, are regions where the red, dusty material has been blown away by winds in the thin atmosphere, exposing the darker rocks below. Not all of them are depressions; Syrtis Major is a relatively lofty plateau.

Life? Well, the famous "canals" do not exist, and were simply tricks of the eye. Conditions on Mars are not suitable for advanced life, and there can be nothing so complicated as a blade of grass; the super-intelligent Martians have been banished to the realm of myth. There may well be very primitive single-celled organisms, but that is all.

Unmanned rockets were first sent to Mars during the 1960s, and by now the whole of the planet has been mapped in detail. It has even been said that we know Mars better than we know some of the most inaccessible regions of Earth. There are craters, valleys, mountains and huge volcanoes, one of which – Olympus Mons – is three times the height of our Everest, and is crowned by a complex, 40-mile caldera. It is usually believed that all the volcanoes are extinct, but we cannot be sure. Certainly Mars was very active volcanically at one stage in its history.

Spacecraft have been landed on Mars, and have sent back information direct from the surface. Plans for manned expeditions are already being discussed, and it seems likely that the first expeditions will be dispatched within the next few decades. We cannot turn Mars into a second Earth, and we cannot give it a breathable atmosphere, but we have to admit that from our point of view it is much more attractive than Venus.

Mars has two tiny satellites, Phobos and Deimos, both discovered in 1877. Neither is as much as twenty miles in diameter, and there is little doubt that they are ex-asteroids, captured from the main asteroid zone in the remote past.

Asteroids may be regarded as cosmic débris. The main belt lies between the orbits of Mars and Jupiter; here we find a whole swarm of midget worlds, not many of which are as much as a hundred miles across. All are airless, sterile and presumably cratered. Several have been imaged from close range by spacecraft passing through the main belt en route to the outer Solar System, and some have bizarre shapes; for example one, Kleopatra, looks remarkably like a dog's bone! Some asteroids, usually very small, swing away from the main zone and pass close to Earth. In February 2001 an unmanned spacecraft, named in honour of the American astronomer Eugene Shoemaker, made a controlled landing on the surface of the asteroid Eros, which is shaped rather like a sausage and has a length of eighteen miles.

Beyond the asteroid zone lie the four giant planets. Jupiter and Saturn may be called gas-giants, while Uranus and Neptune are better described as ice-giants. Their surfaces are gaseous, though no doubt they have solid cores; all four have satellite families, and all four are much more massive than the Earth. Jupiter is brilliant, and Saturn can outshine all the stars apart from Sirius and Canopus, but Uranus is only just visible with the naked eye, and to see Neptune you require optical aid.

Jupiter is the colossus of the Sun's family, and is more massive than all the other planets put together. It is made up chiefly of hydrogen, with a good deal of helium; the bulk of the huge globe – almost 90,000 miles in diameter – is composed of liquid hydrogen. Telescopes show its yellowish, flattened disk, with its dark cloud belts and special features such as the Great Red Spot, a vast whirling storm whose surface area is greater than that of the Earth. Spacecraft have encountered the planet, but manned flight is out of the question, if only because Jupiter is surrounded by belts of radiation which would quickly kill any astronaut unwise enough to venture inside them. There are four major satellites, of which three are larger than our Moon. The fourth, Europa, has an icy surface, below which there may be a sea of liquid water. Of the others, Ganymede and Callisto are icy and cratered, while Io is wildly volcanic.

Saturn, moving far beyond Jupiter, is of the same basic nature, but its system of rings makes it, in my view, the most beautiful object in the entire sky. The rings are made up of icy particles, moving round Saturn as though they were tiny satellites. When the planet is best placed – as during the first years of the present century – even a small telescope will show the ring system well, together with the main atmospheric belts, and occasional white spots. Numerous satellites have been found. One of these, Titan, is larger than the planet Mercury, and has a dense atmosphere made up largely of nitrogen, though the low temperature and the large amount of atmospheric methane means that life there is highly unlikely. We may well find out in November 2004. A spacecraft, Cassini-Huygens (Cassini because the main gap in Saturn's ring system was discovered by the Italian astronomer G.D. Cassini in 1675, and Huygens after the Dutch observer Christiaan Huygens, who discovered Titan in 1655) is scheduled to go there; Cassini will orbit Saturn, and the Huygens probe will, we hope, make a controlled landing on Titan. The spacecraft was launched in 1997, so it has been a long time on the way. Whether Huygens will land on rock or ice, or splash down in a chemical ocean, remains to be seen.

Uranus and Neptune are also giants, but differ markedly from Jupiter and Saturn. They are essentially icy in nature, and have extensive satellite families as well as faint rings. Most of our knowledge about them comes from one spacecraft, Voyager 2, which by-passed Jupiter (1979) and Saturn (1981) before going on to fly by Uranus (1986) and Neptune (1989). Voyager 2 will never return; it is still sending back signals, but before long we are bound to lose touch as it makes its way out of the Solar System. We will never know its eventual fate.

Finally there is Pluto, discovered as recently as 1930 by my old friend Clyde Tombaugh at the Lowell Observatory in Arizona. Pluto is smaller than the Moon, and has a companion, Charon, whose diameter is more than half that of Pluto itself. Interestingly, Charon's orbital period is the same as Pluto's "day", so that to a Plutonian observer Charon would stay motionless in the sky. Whether Pluto should be classed as a true planet is questionable, particularly as it has a very eccentric orbit, and for part of its 248-year revolution period it is actually closer-in than Neptune (as it was between 1979 and 1999). In recent years numerous smaller bodies, of asteroidal size, have been found in this remote part of the Solar System, making up what is called the Kuiper Belt in honour of the Dutch astronomer Gerard Kuiper. It may well be that Pluto is merely the senior member of the Kuiper swarm.

Comets are the most erratic members of the Sun's family. The only reasonably substantial part of a comet is its nucleus, which is never more than a few miles in diameter, and is made up of "rubble" mixed with ice. Most bright comets have very eccentric orbits, and when they are a long way from the Sun they are inert; a comet has been aptly described as a dirty iceball. However, when a comet moves in to perihelion its ices begin to evaporate, so that the comet develops a head or coma and in many cases a tail or tails. All this material is lost, so that by cosmical standards a comet is short-lived. Only one bright comet, Halley's, is seen regularly; it comes back every 76 years, and is due back in 2061. The last really spectacular comet, Hale-Bopp, was a splendid object for months in 1997, but will not return to perihelion for over four thousand years. (Its name honours its co-discoverers, Alan Hale and Tom Bopp.)

All other short-period comets are faint, and very few attain naked-eye visibility. They come from the Kuiper Belt, but brilliant visitors such as Hale-Bopp come from the Oort Cloud, which is several million million miles out in space. If an Oort Cloud comet is perturbed for any reason, it swings inward, and eventually arrives in the inner part of the Solar System. One of several things may then happen. It may simply swing round the Sun and return to the Oort Cloud, not to be seen again for many centuries; it may be trapped by the gravitational pull of a planet, usually Jupiter, and forced into a short-period orbit; it may simply break up and disappear, or it may fall into the Sun and be destroyed. In 1994 one comet, Shoemaker-Levy 9, committed suicide by crashing into Jupiter, causing a disturbance in the Jovian atmosphere which persisted for months.

As a comet moves, it leaves a dusty trail in its wake. When the Earth passes through such a trail, particles dash into the upper atmosphere and are heated by friction against the air-particles, so that they die spectacularly in the streaks of light which we call meteors or shooting stars. What we see, of course, is not the meteor itself, which is of sand-grain size, but the luminous effects it produces during its headlong plunge. Meteors burn out at about 40 miles above sea-level, but now and then the Earth is hit by larger bodies which survive the drop to the ground and land more or less intact, producing craters. These objects are called meteorites. They are not connected with comets or with shooting stars, and come from the asteroid belt. Go to Arizona, not far from the town of Winslow, and you will see a magnificent crater, three-quarters of a mile across, formed about 50,000 years ago – fortunately before anyone lived in the area.

Could we be in danger from a tumbling asteroid, meteoroid or comet? The answer is "yes", and there is a well-supported theory that this did actually happen, some 65 million years ago, causing a change in world climate drastic enough to wipe out the dinosaurs. The chances of a damaging impact in the near future are very slight, but we cannot claim that they are nil. All we can say is that if it does happen, we must hope that we cope with the situation better than the dinosaurs did.

Note, too, that the Solar System contains a vast amount of thinly-spread material. When this material, mainly in or near the main plain of the system, is lit up by the Sun, it causes the cone-shaped glow of the Zodiacal Light. This is not easy to see from light-polluted Britain, but from countries with clearer skies it can be truly beautiful.

I am afraid that this has been a very sketchy tour of the Sun's family, but it did seem to be necessary. Having set the scene, it is time to turn back to Venus.

3

Myth and Legend

Venus is so much brighter than any other object in the sky, apart from the Sun and the Moon, that it must have attracted attention even in the earliest days of mankind. Every race has its folklore and its legends about it. But before going any further, let me try to answer one question which has been asked time and time again during the last two thousand years: can Venus have been the Star of Bethlehem?

The answer is a very emphatic "no". Of course, everyone knows the story, as related in the Gospel according to St Matthew: how the Wise Men from the East crossed the desert, guided by a brilliant star which "went before them" and stopped over the place where the infant Jesus lay. Unfortunately our positive information is very scanty because St Matthew says very little, and none of the other Gospellers mention it at all. Neither are there any records of any unusual astronomical objects seen around that time, and at least we are reasonably confident about our chronology; Jesus was born about 4BC*. (Venus was very familiar, and if the Wise Men could have been taken in by it they would hardly have been very wise.) In fact, the Star cannot have been any ordinary object. It is true that Venus and Jupiter were close together in the sky for a while in 2BC, but the Wise Men would have known about this conjunction, and in any case the time-scale is wrong. If St Matthew's account is to be

---

• Note that there is no Year 0; according to our calendar, 1BC was followed immediately by AD1 – so that the first day of the present millennium was 1 January 2001, not 1 January 2000. Moreover, Christ's birthday was not celebrated on 25 December until several centuries later – by which time the real date had been forgotten, so that our Christmas is wrong too.

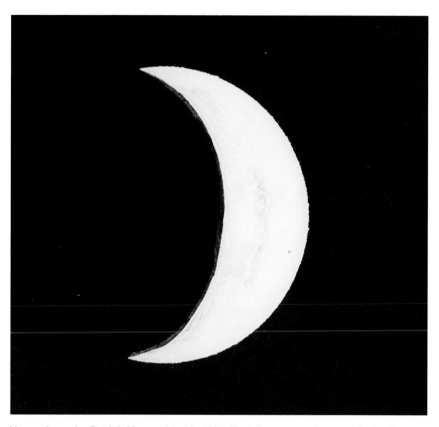

Venus drawn by Patrick Moore, July 18, 1980. The telescope used was a 5-inch refractor. Venus is at its very brightest when it is in the crescent phase.

taken at its face value, then the Star was seen only by the Wise Men, and moved in a most unstellar manner. My own suggestion is that it may have been due to two meteors, which flashed across the sky in the same direction one after the other. This may or may not be the true explanation, and there have been countless theories about the nature of the Star, but since they do not involve Venus they do

To the Greeks and Romans, Venus (or Aphrodite) was the goddess of beauty and love, represented here by the Venus de' Medici, one of the most famous works of art from the classical world. The statue that survives today is a copy from around 100 BC of an original from the 4th century BC – probably by the Greek sculptor Praxiteles.

not concern us here. So let us turn back to really ancient times.

Originally it was assumed that the "Evening Star" and the "Morning Star" were two different bodies. The Chinese called Venus Tai-pe or "the Beautiful White One". In Egypt, Venus or "Bonou", the Bird, was Ouaîti as an evening star and Tioumoutiri as a morning star. (The Egyptian theories about the structure of the universe were rather different from ours; they believed that the sky was formed by the arched body of a goddess with the rather appropriate name of Nut!) The Phoenician name for Venus was "Astarte".

The earliest definite observations of Venus which have come down to us are Babylonian, and are shown on the Venus Tablet discovered by Sir Henry Layard at Konyunjik and now on view in the British Museum. We know that it was written during the reign of King Amisaduqa, and this means that we can date it, because Amisaduqa was on the throne between 1646 and 1626BC. "In

Month XI, 15th day, Venus disappeared in the west. Three days it stayed away, and then on the 18th day became visible in the east. Springs will open, and Adad will bring his rain and Ea his floods. Messages of reconciliation will be sent from King to King'. Adad, by the way, was the weather god of Babylonia and Assyria; Ea was an ancient weather god of Sumerian origin, worshipped also in Babylonia. Note also that the Sumerians were probably the first to master the art of writing; they date back to 5000BC.

Of course these records are part-astronomy, part-astrology; the Tablet was one of a set mercifully preserved in the library of the Assyrian king Ashurbanipal, who reigned from 669 to 626BC, so that it was very ancient even then.

The Chaldaeans, who may well have been the first to divide up the stars into constellations, called Venus "star", the Bright Torch of Heaven – the mother of the gods, and the personification of women. Temples to her were set up at various sites. There is also a Babylonian legend which is very similar to the later and much more famous legend of Ceres and Proserpina. In Babylonia, Ishtar was responsible for the world's fertility, and when she visited the Underworld to search for her dead lover Tammuz all life on earth began to die; it was saved only by the intervention of other gods, who revived Tammuz at something later than the eleventh hour.

Old references to Venus are found everywhere. Homer, in the *Iliad*, calls it "the most beautiful star set in the sky". The Greeks were the first to name it after the Goddess of Beauty; the month of April was regarded as sacred to her, and our name Friday is derived from the Anglo-Saxon "Frigedaeg" (Friga, or Venus, and dae, day). Even the Australian Aborigines had a special name for it: Barnujbir as a morning star, when it was regarded as a special sign for those people who were out hunting at dawn.

The great early New World civilizations were the Maya, the Aztecs and the Inca. All these paid close attention to Venus. Indeed the Maya, of Yucatán, developed an elaborate calendar based upon the movements of the planet. The Maya established a true empire, and we can still admire some of their handiwork, notably the magnificent structure at Chichén Itzá with its Temple of Kukulkán

Chichén Itzá, with its temple of Kukulkán (A. Clarkson).

or Quetzalcoatl, the "Feathered Serpent" god – a 90-foot pyramid built during the eleventh to thirteenth centuries AD directly upon the elaborate foundations of earlier temples. This is the most important ceremonial feature of Chichén Itzá, and is a mine of information about the Mayan calendar; it is, of course, astronomically aligned.

The Maya watched Venus, and found that every eight years it rose as far south in the sky as it can ever do. Since Venus was a major deity, this eight-year period was regarded as particularly significant, and was referred to as the Grand Cycle. Unfortunately we know less about the Maya than we ought to do, because of the activities of one man, Diego da Landa, without doubt the most notorious vandal in the history of science.

The Maya had a somewhat chequered history, ending tragically around 1540 when their country was invaded and conquered by the Spanish. Diego da Landa, a Franciscan priest, was sent to Yucatán ostensibly as a missionary. In the interests of Christianity, he decided to wipe out all traces of Mayan religion and culture, and he was frighteningly efficient. In June 1562 he recorded that

"These people also used certain glyphs or letters in which they wrote down their ancient history and science... We found a great number of these books in Indian characters, and because they contained nothing but superstition and the Devil's falsehoods we burned them all; and this they felt most bitterly and it caused them great grief". Very little survived, though we do have the Dresden Codex, a band of paper describing the results of Mayan observations and calculations of astronomical phenomena. For his outstanding work, da Landa was made Bishop of Yucatán. Other benefits introduced by the Spaniards included cholera, malaria, measles, smallpox and typhoid, so that within a hundred years the Maya population had been reduced by about 90 per cent. Whether the survivors appreciated the advantages of Christianity must be regarded as questionable.

In Mexico the Aztecs built up a civilization which reached its peak during the fifteenth century; the capital, Tenochtitlan, was the centre of a mighty empire. The Aztecs, too, were concerned with astronomical observations; they too were conquered by the Spanish. Tenochtitlan fell in 1519, and again in the interests of Christianity the conquerors set to work to eradicate all traces of Aztec religion. Tenochtitlan was destroyed; the modern Mexico City stands on its site. Finally there were the Inca, who developed a brilliant civilization in parts of what are now Peru and Ecuador. It lasted from about 1438 to 1532, when it fell to the Spaniards – with the usual devastating results. Unlike the Maya, the Inca had no written language but they were first-rate engineers and architects. Astronomically they paid special attention to Venus, both as an evening star and as a morning star.

Whether the Maya, the Aztecs or the Inca made human sacrifices to Venus is not clear, but they may well have done. Venus worship persisted until surprisingly modern times. It has been claimed that in Polynesia human sacrifices were offered to the Morning Star as lately as the nineteenth century, and it has also been said that sacrifices were offered by the Skidi Pawnee Indians of Nebraska. The evidence is highly uncertain, but ancient rituals are slow to die.

Scientific eccentrics have always been with us, and can be genuinely, albeit unintentionally, amusing. They are of many varieties, ranging from the Flat Earthers to the astrologers, the Hollow Globers, the Flying Saucer devotees and so on. Before turning to more weighty matters, it may be worth saying a little about them – beginning with the theories put forward in 1950 by a Russian-born psychoanalyst, Dr Immanuel Velikovsky, who practised in what is now Israel before emigrating to the United States in 1939.

In his book *Worlds in Collision*, Velikovsky claimed that Venus used to be a comet, which was ejected from Jupiter and subsequently bounced about the Solar System rather in the manner of a cosmic table-tennis ball. It made periodical close approaches to the Earth, causing events which are linked with references in the Bible. Thus in 1500BC, at the time of the Israelite Exodus, the comet Venus caused a temporary halt in the Earth's spin, so that the Red Sea was left high and dry for long enough to allow the Israelites to cross. Conveniently, the rotation started up again just in time to swallow up the pursuing Egyptians. The comet Venus returned later on, producing thunder, lightning and other effects noted when Moses was given the Ten Commandments on Mount Sinai. Later encounters produced new phenomena, such as the shaking down of the walls of Jericho. Finally the comet Venus collided with Mars and had its tail chopped off, so that it stopped being a comet and turned into a planet...

What can one make of all this? Dr Velikovsky is an almost perfect example of the pseudo-scientist. Amazingly, his book was taken seriously by some critics, possibly because it had been published by a reputable firm (Macmillan); John J. O'Neill, science editor of the *New York Herald Tribune*, described it as "a magnificent piece of scholarly research", while Ted Thackray, editor of the *New York Compass*, went so far as to compare Velikovsky with Galileo, Newton, Kepler and Einstein.

Subsequently, Velikovsky became something of a cult figure in the United States, and he produced other books following up his original theories, each of which was weirder than the last.

Eventually there was a meeting at which Velikovsky was confronted by various astronomers; the proceedings were published in 1977. All the Velikovskian hypotheses were discussed – such as his claim that during their various encounters in Biblical times Venus, the Earth and Mars exchanged atmospheres, and that these encounters heated Venus to incandescence, so that it is now cooling down quickly enough for the drop in temperature to be measured over periods of a few years.

The trouble is, of course, that Velikovsky's ignorance of astronomy (and, indeed, science in general) was so complete that there was no common ground upon which rational argument could be based. A comet is of extremely low mass by planetary standards – it would take 60,000 million bodies the mass of Halley's Comet to equal the mass of the Earth – and the idea that a comet can change into a planet, or vice versa, is utterly absurd. Neither is there any need to dwell on the mathematical impossibility of a body such as Venus having its orbit abruptly changed from an erratic ellipse into an almost perfect circle, or the equal impossibility of a planet being shot out from a Jovian volcano. The whole episode is interesting psychologically, and of course Velikovsky was himself a qualified psychoanalyst. I met him only once, and found that trying to argue with him was rather like trying to eat tomato soup with a fork. He has at least the distinction of being one of the few scientific crackpots to have attracted official attention – a distinction he shares with Hans Hörbiger in Nazi Germany, who believed that the universe was made up almost entirely of ice, and Trofim Lysenko, who single-handedly stopped the advancement of the science of genetics in Soviet Russia for more than twenty years.

Nowadays we still have the astrologers, and also the flying saucer or UFO cult. I will have more to say about these cults later. Meanwhile, it seems time to return to true science.

4

The Movements of Venus

Venus moves round the Sun in an almost circular path. Its orbital eccentricity is 0.007, less than for any other planet (its nearest rival is Neptune, with 0.009), and the distance from the Sun alters very little, from 66,749,000 miles at perihelion out to 67,760,000 miles at aphelion. On average, Venus receives about twice as much solar radiation as we do on Earth. To us, the Sun has an apparent diameter of about 32 minutes of arc, but from Venus this is increased to over 44 minutes of arc.

Although the orbit is so nearly circular, it is tilted with respect to ours. The angle of inclination is 3 degrees 24 minutes of arc. This may not seem a great deal, but it is more than for any other planet apart from Mercury and Pluto, and means that transits of Venus across the Sun's disk are few and far between. I will have more to say about transits in Chapter 9. Since Venus is closer to the Sun than we are, it moves more quickly; its mean orbital velocity is 21.75 miles per second (78,300 mph) as against 18.5 miles per second (66,500 mph) for the Earth. Travelling at this greater rate in a smaller orbit, Venus takes less time to complete one circuit; the sidereal period or Venus' "year" is only 224 days 16 hours 48 minutes. "Days" here mean Earth-days; Venus itself has a very curious "day", as we shall see later.

The diagram opposite explains why Venus shows phases, from new to full. Needless to say, the diagram is not to scale, and moreover the Earth is taken to be stationary at point E. The fact that it is actually moving round the Sun does not affect the principles involved.

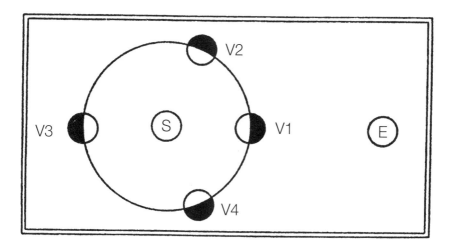

Phases of Venus seen from Earth (E) – at V1, Venus is in inferior conjunction, at V2 in western elongation, at V3 in superior conjunction, and at V4 in eastern elongation.

In position V1, Venus is said to be at inferior conjunction, as it lies between us and the Sun. On the rare occasions when alignment is perfect, the planet is seen as a black disk against the bright body of the Sun, but the orbital inclination means that this does not happen very often. Unless in transit, Venus is invisible at the exact time of inferior conjunction, since its non-sunlit side (blackened in the diagram) is turned in our direction.

As Venus moves on toward V2, a little of the illuminated hemisphere starts to turn toward the Earth. The planet appears in the morning sky as a slender crescent, becoming brighter and brighter as the phase and the elongation from the Sun increase; at V2 the three bodies make up a right angle, and Venus appears as a half-moon. This is known as dichotomy, from two Greek words meaning, literally, "cut in half". Venus is now at its greatest angular distance from the Sun – about 47 degrees – and is a magnificent object in the eastern sky before dawn.

As it moves onward, Venus changes from a half into a three-quarter or "gibbous" phase, and draws back toward the Sun's line of sight, so that it grows steadily less and less conspicuous. By the time it has reached V3 it is full, at superior conjunction, but as it is so close to the Sun in the sky it is difficult to follow. On rare occasions it may even pass directly behind the solar disk, but this phenomenon is, naturally, unobservable. Telescopically, Venus can be found when only a degree or two away from the Sun's limb, and at some superior conjunctions can be kept in view all the time, though such observations are of no real use.

As Venus moves on from V3 towards V4, it appears low down in the evening sky as a gibbous disk, shrinking gradually to half as its angular distance from the Sun increases. It reaches evening (eastern) elongation at V4, and is again at dichotomy, after which it narrows to a crescent as it returns to inferior conjunction at V1. It is obvious that during evening apparitions, Venus is on the wane; at morning (western) apparitions it waxes as it approaches superior conjunction.

The synodic period of Venus, the interval between successive inferior conjunctions, is 584 days, though this is only an average value and may vary by as much as four days either way. In general, about 144 days elapse between an evening and a morning elongation (V4 to V2), while about 440 days are needed for the planet to pass beyond the Sun and come back to evening elongation (V2 to V4).

The minimum distance between Venus and the Earth occurs at the time of inferior conjunction, and may be reduced to as little as 24,600,000 miles – about a hundred times as far away as the Moon, but about ten million miles less than Mars at its closest. Unfortunately it is then invisible, except during a transit. As the phase grows, the distance increases and the apparent diameter shrinks, as shown opposite. The black circle represents the size at inferior conjunction, when the diameter is 65 seconds of arc; the third position shows the apparent size at about the time of greatest

The apparent diameter of Venus is greatest when in the crescent stage, and shrinks as the planet moves toward superior conjunction. These images were all taken on the same scale.

brilliancy (angular distance from the Sun about 40 degrees), and the last position indicates the size at superior conjunction, when the apparent diameter has been reduced to a mere nine and a half seconds of arc, smaller than that of remote Saturn.

Venus is at its best during the crescent stage; the planet is then almost at its closest to us, which more than compensates for the fact that only a relatively small part of the sunlit hemisphere is turned in our direction (the first man to realize this was Edmond Halley). Venus can then cast perceptible shadows. I well remember that some years ago, when I was at the La Silla Observatory in the Atacama Desert of Chile, Venus was superb; I stood against a white wall, and my shadow was very obvious indeed. Mind you, I have to admit that I am rather good at casting shadows!

Keen-sighted people can often see Venus with the naked eye even when the Sun is well above the horizon. This was known in ancient times; Varro records that "in his voyage from Troy to Italy, Aeneas constantly perceived this planet, notwithstanding the presence of the Sun above the horizon". Much more recently, an instance of the daylight visibility of Venus has crept into history. The famous French astronomer François Arago recorded that when Napoleon Bonaparte went to Luxembourg in 1797, to attend a fête given in his honour by the French Directory, he was much surprised to see that the crowd in the Rue de Tournon paid more attention to the sky than to his august self. He was told that although it was noon, the people were watching Venus, which they took to be the guardian star of the Conqueror of Italy; and on looking in the direction indicated, Napoleon saw it for himself. Whether he regarded it as a sign of divine approval is not clear.

Binoculars show the phases easily, and people with really good eyesight can make out the crescent with no optical aid at all. Interestingly, old legends often refer to the "horns" of Venus. Pliny represents Venus as a human figure with two horns; the star-gazers of Samoa also said that Venus is "horned", and so on. The Assyrian Venus is shown bearing a staff tipped with a crescent. To test this sort of observation, I once carried out a rather unkind experiment during a BBC *Sky at Night* television programme. Venus was well placed; I showed a picture of the crescent, and asked viewers to send me sketches of what they could see. In fact the image on the screen showed the view as seen through the telescope. I had many dozens of sketches – and most viewers showed the telescopic view – it is only too easy to "see" what one expects to see. But there were seven drawings, accompanied by puzzled letters, which showed the correct shape, and these were of course sent in by people who had genuinely seen the crescent.

Before pressing on, I cannot resist digressing briefly and delving back into history – because it was the phase cycle of Venus which provided one of the earliest practical proofs that the Earth moves round the Sun.

Early civilizations believed, naturally enough, that the Earth lies at rest in the centre of the universe, with the entire sky rotating around it in a period of 24 hours. There were a few dissentients; around 270BC one of the Greek philosophers, Aristarchus of Samos, was bold enough to claim that the Earth was in orbit round the Sun, but few people believed him, and the Earth-centred or geocentric system lasted for well over a thousand years after his time. It was brought to its highest state of perfection by Ptolemy of Alexandria, who flourished around AD150. We know absolutely nothing about his personality, but we owe him a great deal, because he wrote a book, usually known by its Arab title of the *Almagest*, which was really an encyclopaedia of all the scientific knowledge of the Classical period. The original has been lost, but fortunately the Arab translation has come down to us. The geocentric theory of the universe is always known as the Ptolemaic system, though it was based on earlier work, and Ptolemy simply improved it and summarized it.

The Ptolemaic Earth is central, with the various celestial bodies moving round it in perfectly circular paths. First comes the Moon, obviously the closest object in the sky; then Mercury, Venus and the Sun, followed by the other three planets then known (Mars, Jupiter and Saturn) and finally the stars.

However, difficulties in the way of this arrangement were evident. So far as could be ascertained, the Sun and the Moon moved regularly and steadily across the heavens from west to east, but this was not true of the planets, which were found at times to stand still for a few days against the starry background and then move westward, or "retrograde", before stopping again and then resuming their eastward march. To account for this, Ptolemy supposed that a planet moved in a small circle or "epicycle", the centre of which – the "deferent" – moved round the Earth in a perfect circle. The possibility of elliptical orbits did not occur to him. The circle was assumed to be the perfect form, and nothing short of absolute perfection could be allowed in the heavens.

Mercury and Venus raised problems of their own, and Ptolemy was forced to suppose that their deferents remained permanently in a straight line joining the Sun to the Earth. This did at least explain why Mercury and Venus never appear opposite to the Sun in the sky, but the whole system was clearly artificial and clumsy.

The diagram overleaf shows the movements of Venus according to the Ptolemaic theory. E is the Earth, at rest in the centre of the universe; S is the Sun, D the deferent of Venus, and V1 to V4 the planet in four positions as it moves in its small circle or epicycle. It must be remembered that according to Ptolemy the line EDS is always straight. Since a planet shines only by reflected sunlight, it is obvious that on this scheme Venus can never appear as a full disk, or even a half. At V1 and V3 the dark night-hemisphere will be turned toward us, so that Venus will be invisible; at V2 and V4, part of the daylit side will face us,

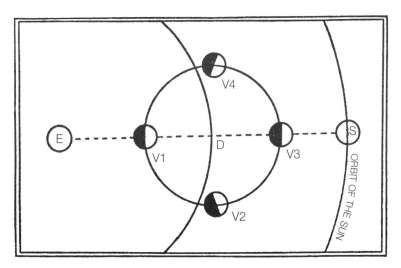

Ptolemy's epicycles, used to explain the movements of Venus in his cosmology.

and the planet will appear as a slender crescent.

There matters rested until the invention of the telescope, in or around 1608. In 1610 Galileo, the first great telescopic observer, turned his newly made "optick tube" toward Venus, and found that there were indeed phases – but not Ptolemaic phases. Sometimes Venus was a crescent, sometimes a miniature half-moon, sometimes almost full.

In those days it was customary to announce discoveries in anagram form. Accordingly Galileo sent a message to Johannes Kepler, the German mathematician who was the first to draw up the Laws of Planetary Motion which bear his name. Galileo's message read:

'*Haec immatura, a me, iam frustra, leguntur – o.y.*'
Translation: '*These things not ripe (for disclosure) are read by me.*'

Rearranged, however, the letters make the following sentence:

'*Cynthia figuras aemulatur Mater Amorum.*'

Or: '*The Mother of the Loves imitates the phases of Cynthia.*'

The Mother of the Loves is, of course, Venus, while Cynthia is the Moon. Galileo had been quick to realize the significance of what he had seen. Venus could not possibly move in the way assumed by Ptolemy. Both it and the Earth had to be planets moving round the Sun.

The Church authorities were not pleased at the idea of demoting the Earth from its proud central position, but Galileo was already a firm believer in the Sun-centred theory which had been proposed by the Polish mathematician Copernicus in 1543. Everything he saw with his telescope convinced him that Ptolemy had been wrong. This is no place to retell the story of how Galileo was brought to trial in Rome by the Inquisition, and forced to "curse, abjure and detest" the false theory that the Earth moves round the Sun, but the facts could hardly be denied. Galileo was finally pardoned by the Vatican – in 1992. Nobody can accuse the Catholic Church of making hasty decisions!

From Venus, the Earth would naturally be an outer or superior planet, and if it could be seen in a cloudless sky, it would appear brighter than Venus ever does to us, because it would remain virtually full. But in fact it could never be seen from the planet's surface; the clouds never clear, and even the Sun would be hidden. There is no such thing as a sunny day on Venus.

5

Venus Through the Telescope

Glorious as it is when seen with the naked eye, Venus is much less spectacular when observed through a telescope. Even a modest instrument will show the polar caps of Mars, the belts and principal satellites of Jupiter, and the rings of Saturn, but Venus appears virtually featureless, and generally nothing definite can be seen apart from the characteristic phase, simply because we are not looking at a solid surface; all we are seeing is the top of an all-concealing layer of cloud. The description given as long ago as 1850 by Sir John Herschel can hardly be bettered:

> *'We see clearly that its surface is not mottled over with permanent spots like the Moon; we notice in it neither mountains nor shadows, but a uniform brightness, in which sometimes we may indeed fancy, or perhaps more than fancy, brighter or obscurer portions, but can seldom or never rest fully satisfied of the fact.'*

Galileo was the first serious telescopic observer. From the winter of 1609–10 he ranged round the entire sky, and, as we have noted, his detection of the phases of Venus was a powerful argument in favour of the Copernican theory rather than the Ptolemaic. But even his most powerful telescope magnified a mere 30 times, and compared with modern instruments it gave poor definition, so that Galileo could hardly hope to see any details on Venus. Neither were his immediate successors more fortunate. Probably the best observer of the mid-seventeenth century was Christiaan Huygens, of pendulum clock fame, who was certainly the first to record

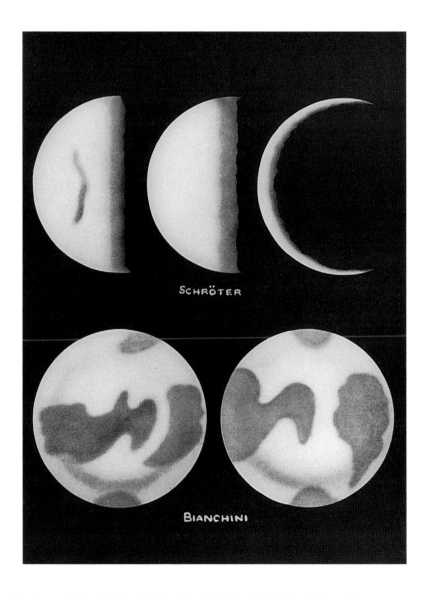

Drawings of Venus by Bianchini (1726) and Schröter (1786). The dark markings are certainly spurious.

detail on Mars; his drawing of the famous dark marking known today as the Syrtis Major was amazingly accurate in view of the small-aperture, long-focus refractor which he used. The Syrtis Major was recorded in 1659. Yet even Huygens failed to make out anything on the brilliant disk of Venus, and this makes one highly suspicious of the positive results claimed by observers of lesser skill.

In 1645 Francesco Fontana, a Neapolitan lawyer and amateur astronomer, recorded a "dark patch in the centre of the disk of Venus". Fontana's telescope was home-made, and probably not much better than Galileo's. Earlier, in 1636 and 1638, he had made similar sketches of Mars, showing a circular disk with a ring inside it and a dark patch in the centre. There is no doubt that these effects, with both Venus and Mars, were purely optical. Fontana was followed in 1665 by another Italian amateur, Burattini, who saw the same kind of thing – again purely optical.

Next, in 1667, came Giovanni Domenico Cassini, who made his observations from Bologna. He recorded various bright and dusky patches on Venus; his first sketch was made on 14 October 1666, at 17 hours 45 minutes, and from observations made during this period he produced the first estimated rotation period – 23 hours 21 minutes, only a little shorter than that of the Earth. Subsequently Cassini left Italy to become the first Director of the newly founded Paris Observatory, and in the less transparent skies of France he was unable to recover the markings on Venus. His son J. J. Cassini, who followed him as Director, was almost equally unsuccessful, though he did try to confirm the rotation period.

The elder Cassini had several important discoveries to his credit, including the main division in Saturn's ring system; he also found four of Saturn's satellites (Iapetus, Rhea, Dione and Tethys). He was unquestionably an excellent observer, but it is not easy to understand why he so completely failed to recover the markings on Venus after 1667 – if he had ever seen them in reality. He was much more successful in determining the rotation period of Mars, giving a value of 24 hours 40 minutes, which is less than three minutes too long. It may be unfair to dismiss his Venus shadings as spurious,

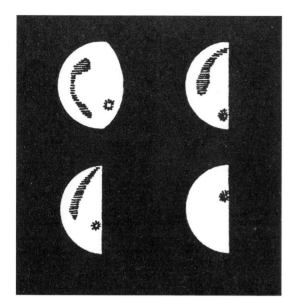

Drawings of Venus by G. D. Cassini (1667).

but the planet is a particularly difficult object to study with a small refractor, which is bound to give a good deal of false colour. The best we can say is that his observations of markings on Venus are of questionable validity.

The next positive observations were made more than half a century later by Francesco Bianchini, in Rome. In 1726 he began a study of Venus with a refractor of 2.5 inches aperture and 66 feet focal length, using a magnification of slightly over 100. Bianchini's results were decidedly startling. They included a map, shown here, with "oceans" and "continents". From his drawings, he concluded that the rotation period was about 24 days.

It is worth saying a little more about Bianchini, because although his "markings" on Venus were unquestionably spurious, he was both energetic and resourceful. He was born in 1662, and became librarian to the Pope, Alexander VIII, in 1684; subsequently he was promoted to the office of papal chamberlain. His papal connections meant that he could set up his observing equipment on the Palatine Hill, and in 1725 he did so, but at first things did not go quite according to plan:

'But my astronomical efforts were nearly thwarted by my study of Antiquity. For on the next day, August 17th, I was on the part of the Palatine Hill which looks towards the church of San Gregorio Magno to the south-east on the slope of Scauri. While I was exploring the ruins of the Palace of the Caesars to investigate its layout from the surviving walls, (the discovery at that time of an extensive basilica, and the chambers belonging to it in the Farnese gardens provided the opportunity for this investigation. The grandeur of its structure and ornamentation has revealed to the eyes of the citizens of Rome the magnificence of the ancient Emperors, an example of which I will provide, God willing, when I have worked out the ground plan), and while I was carelessly running about to take the measurements of the surviving rooms in the east wing of Augustus' home, whose ruined walls survive in the vineyard of the English College, I fell into a broad, deep hole in the pavement which I had not noticed as I rushed around with my eyes fixed on the point to which I was going to measure; I broke my right thigh, and by God's mercy was only saved from the imminent death which threatened me by pressing with all my power with both hands and my left foot against the sides of the hole, to sustain the weight of my body and avoid falling headlong into the pit forty palms below, the depth of which I knew from my measurement of the lower chamber. The injury to the thigh interrupted the observations I had begun, which however with God's grace I was able to resume even more successfully at the start of the following year, 1726.'

Bianchini made some very reasonable drawings of features on the Moon (incidentally, he also discovered a couple of comets). He then turned his attention to the shadings on Venus:*

'After the selection of these observing sites, therefore, whenever we had the opportunity in February and March we made daily notes of

---

* Bianchini's book was published in 1728, naturally in Latin. It remained untranslated until 1999, when, at my instigation, an English version was produced by Sally Beaumont, and published by Springer Verlag under the title *Observations Concerning the Planet Venus.*

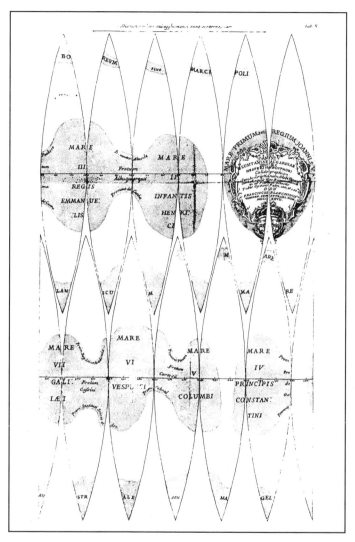

Bianchini's Map of Venus, showing what he believed to be seas and continents similar to those on Earth

the markings which appeared on Venus' globe. They were similar to the larger lunar ones which can be seen with the naked eye on that luminary, called "maria" by astronomers on the "Selonographia" map, for example Mare Crisium, Mare Serenitatis etc. They are really areas on the surface on the globe which are less efficient at reflecting the bright sunlight.

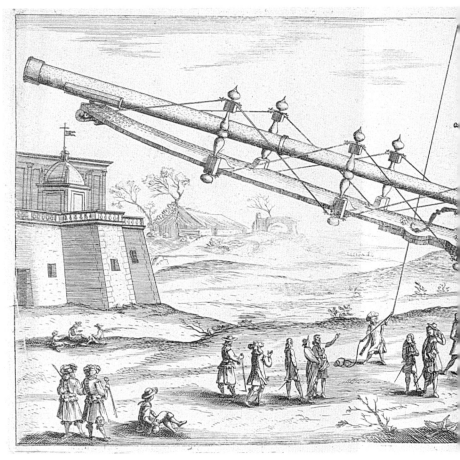

Bianchini's telescope was a refractor, or lens-based design, built for him by Giuseppe Campani. As this engraving shows, it was a monster of its time, almost 70 ft long.

'In order to see these markings more clearly it is necessary to choose days free from mist, but also to wait until twilight is more advanced, at least half an hour after sunset. For the naked eye features called "maria" on the Moon appear to be surrounded only by a washed-out pallor if observed at sunset, but stand out more clearly half an hour later in the darker sky; and in the same way the brighter regions of Venus show in the telescope a much greater contrast with the paler parts of that disk if atmospheric glare is diminished and does not distract and dazzle the eye.'

Bianchini was confident that the markings were real, and that he

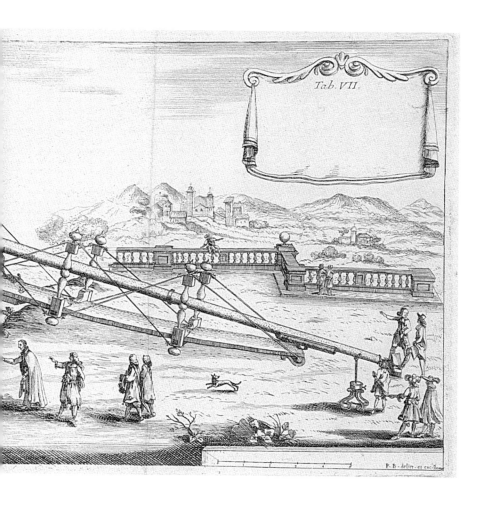

Tab. VII.

could measure the planet's rotation period by watching their drift. He also believed that there were lands and seas, and he gave them romantic names: the Royal Sea of King John, the Sea of Prince Constantine, the Strait of Vasco da Gama, the Sea of Marco Polo and so on. Sadly, we must agree that the markings were purely optical effects, so that his map is of historical interest only, but at least it was a praiseworthy first attempt.

Apart from Galileo's discovery of the phases, the first really useful telescopic result was due to M. V. Lomonosov, who is – rightly – regarded as the first great Russian astronomer. From his modest observatory in St. Petersburg he watched the transit of Venus in 1761, when the planet passed across the face of the Sun.

Lomonosov found that the edge of the disk was slightly blurred and fuzzy, and from this he concluded that Venus is surrounded by "a considerable atmospheric mantle, equal to, if not greater than, that which envelops our own earthly sphere". He was of course quite correct, and the credit for the discovery of Venus" atmosphere must go to him. Rather surprisingly, his announcement attracted little attention at the time, and his great French contemporary Lalande was inclined to believe that Venus had no atmosphere comparable with ours, but the existence of an atmosphere was virtually proved later in the eighteenth century by Johann Hieronymus Schröter.

Schröter has an honoured place in the history of astronomy. He was an amateur who set up his observatory at Lilienthal, near Bremen (where he was Chief Magistrate), and equipped it with the best telescopes he could obtain. He observed the Moon and planets systematically from 1778 until 1814, when his observatory was destroyed by the invading French army and his brass-tubed telescopes were plundered by the soldiers – who mistook them for gold. Schröter was the first really great observer of the Moon, and it may be said that he laid the foundations of selenography. He was equally energetic in observing the surface features of Mars, though admittedly he misinterpreted them; he was also deeply involved in the hunt for the "missing planet" between the orbits of Mars and Jupiter, and indeed the third asteroid, Juno, was found by his assistant, Karl Harding, from Lilienthal. So far as Venus is concerned, his main results were contained in his book entitled *Aphroditographische Fragmente*, published in 1796.

It has often been claimed that Schröter's work was of poor quality. This is emphatically not true, and he seldom made a serious error. His largest telescope – a 19-inch reflector made by Schräder of Kiel – may have been mediocre, but two of his other reflectors were made by William Herschel, and were undoubtedly excellent.

Schröter did not find Venus an easy object. Between 1779 and 1788 he failed to detect any markings at all, but on 28 February 1788 he "perceived the ordinarily uniform brightness of the planet's disk to be marbled by a filmy streak". Subsequently, he saw other markings, but all were diffuse and ill-defined, so that he came to the correct conclusion that they were atmospheric in nature.

There were other confirmatory observations, as Schröter noted. The brightness of the disk falls away perceptibly toward the terminator, which indicates absorption in an atmosphere; also, the horns of the crescent Venus are often seen to be prolonged beyond the semi-circle, which can never be the case with a planet which is virtually airless (Mercury, for instance). Schröter's reasoning was quite sound. The fading of the brightness toward the terminator is almost always very obvious, and the extension of the horns can occasionally be so marked that it may stretch right round the dark hemisphere, giving the appearance of a ring.

All this shows that the atmosphere is of considerable depth. There is also a curious phenomenon connected with dichotomy, i.e. the exact amount of half-phase. When at its greatest elongation, east or west, Venus should of course be seen as a perfect half. The times of the elongation are known very accurately, and it should therefore be possible to predict dichotomy almost to the nearest minute – yet the fact remains that these predictions are almost always wrong. At western elongation, when Venus is a morning object and therefore waxing, dichotomy

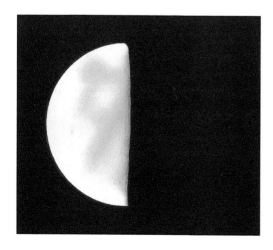

Venus at dichotomy.

Drawing by Paul Doherty,

15-inch reflector x 360.

is invariably late; at eastern elongation, when Venus is waning in the evening sky, dichotomy is early.

This behaviour was first noted by Schröter, and in 1957 I christened it the "Schröter effect", a term which has become generally accepted. In August 1793 Schröter himself found that theoretical dichotomy differed from the observed date by eight days, and the phenomenon was also evident at later elongations. Forty years later the German astronomers Wilhelm Beer and Johann von Mädler, who were the first to draw really accurate maps of the Moon and Mars, found that the average discrepancy was six days, observed dichotomy being early for eastern elongations and late for western. I have studied Venus since 1934, and I have found that the discrepancy can be anything between two and twelve days. The average value, from my work, is about four days.

Estimates of the exact moment of dichotomy are not easy to make, and generally the terminator appears sensibly straight for several days, but the Schröter effect is so pronounced that its reality cannot be questioned. It is certainly due to the dense atmosphere, and I have never found any comparable effect with Mercury.

Schröter was also concerned with trying to measure the axial rotation period of Venus. With Mars there is no problem; simply watch the dark markings as they are carried across the disk by the axial spin. Not so with Venus, because the markings are so ill-defined, and in any case shift and change, but it was the only method available to Schröter, and he did his best. In 1789 he gave a value of 23h 21m 19s, and in 1811 he modified this to 23h 21m 7.977s. Giving a value to a thousandth of a second seems rather ambitious, but it is only fair to add that Schröter was well aware of the problem, as he commented in 1792:

> 'The circumstance also that there are seen on this planet none of the flat spherical forms as are conspicuous on Jupiter and Saturn, none of the strips or longitudinal spots parallel to the equator which are seen on these planets ... and which point out a certain stretch of atmosphere, give room to infer that the globe of Venus ... performs its rotation round its axis in a much longer space of time than these

William Herschel's drawing of Venus as he saw it on 19th June, 1790

planets ... and this is actually confirmed by my observations of the distinct part of Venus.'

The true rotation period of Venus was not found until 1964, and the result turned out to be very unexpected indeed. I will have more to say about it below. Meanwhile, let us turn to a very interesting episode in the story of the observation of Venus: the verbal battle between Schröter and William Herschel concerning the existence or non-existence of high mountains on the surface.

Herschel was of course much the greatest observer of his time. He began his astronomical work in 1772, and in 1781 became world-famous for his discovery of the planet Uranus. It is true that he was not looking for a new planet, and did not recognize its true nature even when he saw it (he mistook it for a comet). It is also true that his main work was in connection with the stars; he was the first to give a reasonably good picture of the shape of the Galaxy, he proved the existence of physically associated pairs of stars (binary systems), and he discovered large numbers of star-clusters and nebulae. He was by far the best telescope-maker of the age, and his largest reflector, completed in 1789, had an aperture of 49 inches and a focal length of 40 feet, though admittedly most of his best work was carried out with smaller telescopes. His observations of the planets were more or less incidental, and confined mainly to the earlier part of his career, but he did pay some attention to Venus. A typical drawing is given here, which Herschel described as follows:

'June 19th, 1780. There is on Venus a bluish, darkish spot ado; *and another, which is rather bright,* ceb; *they meet at an angle, the place of which is about* ¹/₂ *the diameter of Venus from the cusp, a. June 21, 23, 24, 25, 26, 28, 29, 30: Continued observations were made on these and other faint spots ... The instrument used was a 20-foot Newtonian reflector, furnished with no less than five different object specula, some of which were in the highest perfection of figure and polish; the power generally 300 and 450. But the result of them would not give me the time of rotation of Venus. For the spots often assumed the appearances of optical deceptions, such as might arise from prismatic affections; and I was unwilling to lay any stress upon the motion of spots, that either were extremely faint and changeable, or whose situation could not be precisely ascertained. However, that Venus has a motion on an axis cannot be doubted, from these observations; and that she has an atmosphere is evident, from the changes I took notice of, which surely cannot be on the solid body of the planet.'*

Schröter had two telescopes which he had bought from Herschel (incidentally, of all the telescopes which Herschel made it seems that only those which he retained, and the two sold to Schröter, were ever used for serious research). Generally, the two men were

An 'enlightened mountain' sketched by
Schröter above the cusp of Venus

on excellent terms. Their one brush came over the "mountains on Venus" which Schröter believed he had seen.

As we have noted, Schröter had to wait for a long time before seeing any detail on Venus:

> *'I perceived neither spots, nor any other remarkable appearance except the unusually quick decrease of light toward the boundary of illumination, which itself was not sharply defined.'*

Then, on 28 December 1789, using magnifications of from 161 to 370 on his 7-foot focus Herschel telescope, he saw that the southern cusp was blunted, while beyond it there was a small luminous speck. He saw it again on 31 January 1790, and again, three times, in December 1791. By now he was fully convinced that it was an "enlightened mountain" of very considerable height, catching the rays of the Sun – as does actually happen with the Moon, whose peaks near sunrise or sunset over the lunar surface can often be seen apparently well clear of the terminator.

Over the next few years Schröter made further observations of the same kind, and estimated the heights of the peaks to be between 10 and 19 miles. He wrote:

> *'Considering the immense height of the mountains, it is natural to suppose that the shadows of mountains will, no doubt, at times occasion an uneven, ragged appearance...'*

He noted that the terminator – the boundary between the sunlit and night hemispheres of the planet – often seemed jagged rather than smooth, and continued:

> *'Nature seems to have raised on Venus great inequalities, and mountains of such enormous height, as to exceed four, five and even six times the perpendicular elevation of Cimbiraco, the highest of our mountains.'*

At this point Herschel entered the discussion, with a tone quite uncharacteristic of him. He had been making observations of Venus without seeing anything really definite, and in his paper criticising Schröter, he asked: "by what accident I came to overlook mountains in this planet which are said to be of enormous height". He attributed the effects seen by Schröter along the terminator to "a deception arising from undulations in the air", and even questioned the condition of Schröter's telescope, which Herschel himself had made. "Probably the mirror, which was a very excellent one, was by that time considerably tarnished, and had lost much of the light necessary to show the extent of the cusps in their full brilliancy." He stated: "As to the mountains in Venus, I may venture to say that no eye, which is not considerably better than mine, or assisted by much better instruments, will ever get a sight of them."

This was certainly uncompromising. Fortunately Schröter did not rise to the bait, and his reply was calm and courteous: he alluded to "the friendly sentiments which the author has always hitherto expressed toward me, and which I hold extremely precious." Very reasonably, he pointed out that Herschel's main attention was devoted more to stellar observations than planetary ones:

'I should indeed be surprised that the celebrated author had not, in all the time since 1777, perceived any inequality in the boundary of light, or other appearance of that kind, tending to confirm the existence of very high mountains according to the old observations, were it not that his bold spirit of investigation has been chiefly employed in making much more extensive discoveries in the far distant regions of the heavens, where he has gathered unfading laurels. In fact, the observations which he had communicated from his journals are much too few *to prove a negative against old and recent astronomers.'*

Finally, he pointed out that he had never actually seen mountains in Venus, but "had only *deduced* their existence and height from the observed appearances. It is impossible to see them, according to what I have expressly asserted in my paper on the Twilight of Venus."

Schröter's attitude effectively de-fused a potentially explosive situation. So far as he and Herschel were concerned, the matter ended there; neither is there any hint that the two were subsequently upon anything but the best of terms.

Come back now to the problem of the axial rotation period. The first step was to locate the poles of the planet, and this was by no means easy – as it is with Mars, where the poles are covered with white ice-caps which wax and wane with the Martian seasons. It is undeniable that the cusps of Venus often show bright caps or hoods; they were first definitely recorded in 1813, by the German astronomer Franz von Paula Gruithuisen, though they are indicated on some of Schröter's sketches. They can be striking. I have been following them ever since 1934, and they are generally present. But were they truly polar? I was always convinced that they did indeed mark the polar regions, and that they were due to the atmospheric circulation in those areas. This is now known to be correct, but some early interpretations of them were very wide of the mark. It was even suggested that they might be due to lofty, snow-covered plateaux poking out above the cloudy atmosphere; during the latter part of the nineteenth century Schröter's mountain theory was still very popular. The French observer Etienne Trouvelot wrote in 1878 that "the solar patches are distinctly visible … the surface is irregular and seems like a confused mass of luminous points … The surface is undoubtedly very broken, and resembles that of a mountainous district studded with numerous peaks … the polar spots seem to be bristling with peaks and needles".

Almost every observer of Venus attempted to measure the rotation period by following the drifting of the shadings and bright patches, but the results were wildly discordant. Between 1666 and 1962 over a hundred estimates were published – and every one of them turned out to be wrong. The only thing that was *not* suspected was a period longer than Venus' "year", and yet this proved to be the truth. Venus spins in a period of 243 Earth days, and does so from east to west, not west to east as with the Earth and Mars.

Most of the early estimates were of the order of 24 hours, and some of them were taken to an absurd degree of accuracy. In 1895 Leo Brenner, at the Lussinpiccolo Observatory in Italy, gave a value of 23h 57m 36.2396s, amending it in the following year to 23h 57m 36.27728s. One cannot help feeling that this is rather like giving the age of the Earth to the nearest minute. (As a matter of fact, Brenner – an assumed name – was very much of an astronomical charlatan, and the reliability of any of his published work is, to put it mildly, dubious.)

Rather earlier, G. V. Schiaparelli, the Italian astronomer who will always be remembered for his drawings of the canals of Mars, produced an astronomical bombshell in the claim of a rotation period for Venus of 224d 16h 48m, which is precisely the same time taken for Venus to complete one orbit round the Sun. In other words, Schiaparelli came to the conclusion that Venus keeps the same hemisphere turned permanently sunward.

Schiaparelli's main observations of Venus were made from Milan in 1877–8, with a fine 9-inch refractor. As with Mercury, he observed in daylight, so that Venus was high in the sky. From his studies of the southern cusp-cap, he derived what he believed to be a reliable rotation period, and in fact it did not seem to be really surprising; the Moon behaves in the same way relative to the Earth, and so do the principal satellites of other planets with respect to their primaries. Tidal friction over the ages is responsible, and the idea of Venus having a similarly captured or synchronous rotation was reasonable enough. Schiaparelli also studied the elusive surface markings on Mercury, and came to a similar conclusion – a rotation period equal to the orbital period; in the case of Mercury, 88 Earth-days. His opinions carried a great deal of weight, and for Mercury the synchronous rotation was still accepted until radar work in the early 1960s showed it to be wrong (the true value is 58.56 Earth-days). Venus was very much of a problem. Most observers agreed with Schiaparelli, but others did not.

One very strong supporter of Schiaparelli's result was the American astronomer Percival Lowell. I must digress briefly here, because Lowell's views about both Mars and Venus caused a tremendous amount of argument less than a hundred years ago.

Lowell was a man of immense energy. In his early career he had close connections with the Far East, and became an authority on Japanese and Korean culture, but later he gave up most of his other activities in order to devote his time to astronomy. There can be no doubt of his ability, and he was an excellent mathematician; it was his work which led to the discovery of the ninth planet, Pluto, and even if this was fortuitous (as has been claimed) it was still a brilliant investigation. Perhaps unfortunately, Lowell's name

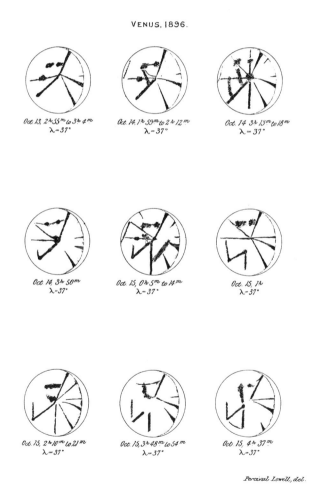

Drawings of Venus, by Percival Lowell. The canal-like features are purely illusory.

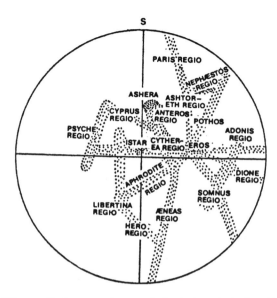

Lowell's map of Venus, with the names he gave to the features he believed he had seen.

will always be associated with the canals of Mars, which I must discuss here because there is a strong link with Lowell's drawings of Venus.

The Martian canals were first described in detail in 1877, by Schiaparelli, who saw them as strange, artificial-looking lines running across the ochre deserts and forming a planet-wide network. The name "canal" implies an artificial origin. Schiaparelli kept an open mind about this, but he certainly thought that the features were cracks through which water flowed equatorwards from the melting polar ice-caps. From 1885 onwards canals became thoroughly fashionable, and Lowell was so enthusiastic about them that he erected a private observatory at Flagstaff, in the clear skies of Arizona, principally to observe them. Between 1894 and his death, in 1916, Lowell recorded hundreds of Martian canals, and was firm in his belief that they were constructed by brilliant engineers in an attempt to defeat the growing shortage of water by a tremendous irrigation scheme.

There was nothing wrong with Lowell's telescope, a superb 24-inch refractor made by Alvan Clark. I know it well, because during my Moon-mapping days, before the Apollo missions, I made extensive use of it. Unfortunately Lowell was not a good observer, and other astronomers, using equally large telescopes, either failed to see the canals at all or else recorded them as nothing more than blurred, hazy streaks. Moreover, Lowell drew canal-like features not only upon Mars, but also upon other bodies, such as Mercury, Venus and the satellites of Jupiter.

His chart of Venus, published in 1897 with an accompanying paper, is a case in point. The markings shown and named by Lowell are definite enough, and the overall effect is most un-Venuslike. From a dark patch, Eros, a sort of focal centre, Lowell drew dark strips which he named Adonis Regio, Aeneas Regio and so on, while there were other less conspicuous streak centres such as Somnus Regio and Ashera. Lowell was firmly convinced that the markings were unchanging, and that the 225-day rotation period was correct, so that the same hemisphere was permanently sunlit. His description of the markings runs as follows:

'The markings themselves are long and narrow; but unlike the finer markings on Mars, they have the appearance of being natural, not artificial. They are not only permanent, but permanently visible whenever our own atmospheric conditions are not so poor as to obliterate all detail on the disk. They are thus evidently not cloud-hidden at any time … There is no distinctive colour in any part of the planet other than its general brilliant straw-coloured hue. The markings, which are of a straw-coloured grey, bear the look of being ground or rock, and it is presumable from this that we see simply barren rock or sand weathered by aeons of exposure to the Sun. The markings are perfectly distinct and unmistakable, and conclusive as to the planet's period of rotation. There is no certain evidence of any polar caps.'

Lowell thus rejected the whole idea of a dense, all-concealing atmosphere, and returned to Bianchini's theory that the markings are permanent features of a solid surface. Incidentally, Lowell seldom took full advantage of the 24-inch refractor, and usually stopped it down to 18 inches, which seems to indicate either that the instrument was not functioning properly or else that there was something radically wrong with the eyes of those who were using it. There is never any need to reduce the aperture of a large telescope when seeing conditions are good, and when conditions are poor there is no point in observing at all. That was the opinion of men such as Edward Emerson Barnard, who worked with the Lick 36-inch refractor, and Eugenios Antoniadi, who used the Meudon 33-inch refractor at Paris. I have observed with most of the world's largest refractors, and I have never had the slightest inclination to stop them down. Neither have I seen canals on Mars, Venus or anywhere else.

Barnard, with the 36-inch, did not in the least agree with Lowell's remark that the features on Venus looked like "steel engravings". He wrote as follows:

'Venus has been examined on a number of occasions with the 36-inch, when the planet was beautifully defined ... Nothing was seen of the singular system of narrow dark lines shown in recent years by some observers to cover the surface of the planet. Every effort was made to show them, by reduction of aperture and by the use of solar screens and various magnifying powers. They were also looked for with the 4-inch finder. Previous attempts with the 12-inch here also failed.'

The streaks on Venus have never been seen through large telescopes, except by Lowell, and there is absolutely no doubt that they were tricks of the eye. The same is true of streaky features recorded by observers using smaller instruments. Most of these show a central dark patch from which lines radiate outwards like the spokes of a wheel, at once indicating an optical effect.

Lowell's "canals" marked the end of one period in the story of the observations of Venus. With the twentieth century, new techniques were developed, and all ideas about the nature of Venus were drastically modified.

6

Telescope, Spectroscope and Speculation

Until well into the twentieth century, almost all of our meagre knowledge of Venus was obtained by visual observation at the eye-end of a telescope. Useful photographs were taken from 1923, and the first really good spectrographic work was carried out in 1932, when carbon dioxide was identified in the planet's atmosphere; temperatures were measured by using thermocouples, and the first radar contact dates from 1961. But it was not until the flight of the first Venus spacecraft, Mariner 2 in 1963, that we had any truly reliable information. Up to that time, Venus was still the "planet of mystery".

Telescopically, the main attention was concentrated upon the dusky shadings, the bright regions and the cusp-caps, and efforts were made to discover the axial rotation period. Large telescopes produced very little in the way of positive results, and it is worth quoting a comment made in 1897 by E. E. Barnard, who was renowned for his keen eyesight and who was using the Lick 36-inch refractor, the second most powerful in the world:

'Surface markings were always present, but they were always very elusive, and at no time could satisfactory drawings be secured … I am confident that the faint elusive spots seen with the great telescope were real, but whether they were of a permanent nature it was impossible to tell, for the same spots were not recognized with certainty at different observations. The impression, however, was that they were not permanent.'

An early photograph of Venus, taken by E. E. Barnard in 1899, 12-inch Equatorial.

Frankly, this description might just as well have been written by an amateur using a home-made telescope. But Barnard's results were confirmed elsewhere, particularly by E. M. Antoniadi, a Greek astronomer who spent much of his life in France, and made use of the fine 33-inch refractor at Meudon, near Paris. He was probably the best observer of the time (for example, his map of Mars proved to be amazingly accurate), and between 1900 and 1939 he paid close attention to Venus, charting the bright and dusky areas as precisely as he could. Yet he always supported Schiaparelli's theory of synchronous rotation.

I have had the advantage of being able to observe Venus with very large refractors, including those at Flagstaff and Meudon, and I have never seen anything really definite, but now and then exceptionally well-defined markings were reported by observers using smaller telescopes. One of these was W. H. Steavenson, a medical doctor who became accepted as a very fine observer, and who achieved the distinction of being one of the few modern amateurs to have served as President of the Royal Astronomical Society. In 1924, using a power of 280 on a 6-inch refractor, he wrote:

'I at once saw a marking which was much more prominent than any I have seen before. Its most conspicuous position took the form of a broad dusky band stretching westwards towards the limb from just south of what would be the centre of the true disk ... It should be visible in any telescope over 3 inches aperture.'

Features as conspicuous as this might be expected to yield a rotation period, but again the evidence was conflicting. In 1921 W. H. Pickering, from visual studies, announced a period of 2 days 20 hours. This was supported in 1924 by Henry McEwen, who devoted a lifetime to the study of the planet.* McEwen, indeed, regarded the problem as definitely solved, and wrote:

'It is due to the skill and acumen of Professor Pickering that we owe the discovery of this unique rotation period ... which will be appreciated at its due value by a future generation of astronomers.'

Unfortunately the "discovery" was not nearly so conclusive as McEwen thought. Steavenson's results were completely different. His favoured rotation period was eight days, or perhaps a sub-multiple of that amount. However, Pickering, McEwen and Steavenson agreed on one point; they believed that the axial inclination of Venus must be very different from that of the Earth, and Steavenson went so far as to suggest that there were times when the poles of the planet lay in the centre of the disk.

This is worth following up, because we do now know the answer. The Earth's axis is tilted to the perpendicular of the orbit by $23^1/_2$

---

* He was Director of the Mercury and Venus Section of the British Astronomical Association for over sixty years; he finally retired in 1957, when I succeeded him.

degrees, which of course is why we have our seasons. The tilt of Mars is practically the same, and those of Saturn and Neptune slightly greater, while Mercury and Jupiter are almost "upright". Uranus, the remote ice-giant discovered by William Herschel, is different; the axial inclination is 98 degrees, giving rise to a very curious calendar (the rotation period is 17.2 hours). Uranus is 31,000 miles in diameter, but could Venus have the same sort of inclination? If so, why?

Actually, the inclination of Venus is over 177 degrees, so that relative to Earth the planet is "upside down", spinning from east to west. Why this is so remains a puzzle. The favoured theory is that at an early stage in its evolution Venus was hit by a massive impactor and literally knocked over. This does not sound at all plausible (any more than it does with Uranus), but it is not easy to think of anything better. At least the cusp-caps do overlie the polar regions, as I always maintained.*

Before the Space Age there were two widely different representations of the elusive markings seen on Venus. In 1926 Steavenson suggested that the polar caps were due to afternoon mist, while Antoniadi dismissed them as mere contrast effects. McEwen wrote that "Venus' atmosphere may be transparent enough to show markings situated (probably) in the lower and denser part of the atmosphere which corresponds to, say, the Earth's troposphere. This may be due to volcanic smoke or dust, or to seeing dimly contrasted features which are on the surface of the planet". Audouin Dollfus, the eminent French planetary observer, believed that the dark, drifting, lower-lying clouds could sometimes clear sufficiently to show the surface features below, while Harold Urey, a Noble Laureate for chemistry, wrote that "the

---

* *En passant* it is interesting to work out the positions of the celestial poles of Venus: for this I am indebted to Iain Nicolson. From Venus, of course, the clouds will always obscure the sky. The north celestial pole (that is to say, the pole lying north of the ecliptic plane), lies at RA 18h 11m, declination +67 degrees 10', between the stars Delta and Zeta Draconis; the south celestial pole is at RA 06h 11m, declination -67 degrees 10'. As with Earth, there is no bright south polar star; the nearest naked-eye candidate is Eta Doradûs, magnitude 5.7. However, Canopus, the second brightest star in the entire sky, is reasonably close to the south celestial pole of Venus.

dark areas observed by Dollfus could be low-lying land awash with seas. In this case they may be observable either as bare, low, flat land or as such land covered with vegetation". There were also sceptics, such as myself, who maintained that the dark shadings and bright areas were nothing more than cloud phenomena in the upper atmosphere of Venus. I was not surprised when attempts to draw up permanent maps of the planet were signally unsuccessful.

Earth-based photography was not initially very helpful, because Venus is a particularly difficult object to photograph well. This is partly because of its dazzling brightness, but also because we cannot hope to record anything but the upper clouds, which are not noticeably informative. However, it was expected that better results could be obtained by taking photographs in the light of certain selected wavelengths. Long waves (red and yellow) are penetrative, while short waves (blue and violet) are not. This, of course, accounts for the blueness of our day sky; the short waves from the Sun are scattered around by the atmospheric particles, while the red and yellow waves can pass through more or less unhindered. When Mars is photographed in red light, the surface

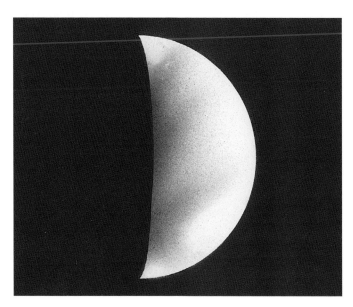

Venus, May 1985 by Paul Doherty, 15-inch reflector x 400.

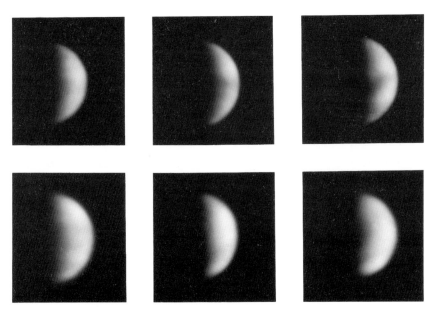

Six photographs of Venus, Mount Wilson 100-inch reflector.

details are clearly shown, whereas in violet photographs no surface detail can be seen – since the short wavelengths fail to penetrate even the tenuous air of Mars, and record only the upper layers of the Martian atmosphere.

It was naturally expected that red light would penetrate the atmosphere of Venus in a similar way, but this hope proved to be unfounded; the red images were entirely blank. Then, in 1927, F. E. Ross at Mount Wilson experimented by taking photographs not in red light, but in ultra-violet. At once, distinct markings were seen. The pictures registered hazy, cloud-like formations as well as darker patches. This was quite contrary to expectation, and, incidentally it dealt a final death-blow to Lowell's theory of a visible solid surface. The markings could only be atmospheric; moreover, they had to be phenomena of the upper clouds, since the ultra-violet rays were unable to penetrate to the lower layers. Some vague streaks were recorded, but to try to correlate them with Lowell's "canals" was obviously absurd.

As soon as Ross' pictures became available, new attacks were made on the problem of the rotation period, but the clouds changed in form and appearance so rapidly that little could be learned from them. Ross derived a very uncertain period of about 30 days, while Dollfus, using photographs taken with the 24-inch refractor at the high altitude Pic du Midi Observatory in the French Pyrenees, found that the rotation must be "very slow". Gerard Kuiper, at the McDonald Observatory in Texas, used the large reflector there to photograph Venus, and derived a rotation period of "a few weeks". It was all very nebulous, and for once the camera seemed to be a broken reed. The next attempts were made by means of spectroscopy.

Just as a telescope collects light, so a spectroscope splits it up. There are 92 naturally occurring fundamental substances or elements, from which all the matter in the universe is made, and each element has its own way of behaving. When a spectroscope is turned toward the Sun, we see a bright continuous rainbow of colour crossed by dark lines which are characteristic of various elements, and by studying the positions of these lines we can find out what substances are responsible for them. Since Venus shines by reflected sunlight, it yields a basically solar-type spectrum, but the dense atmosphere of the planet imprints its own trademarks. Moreover, the spectroscope held out hope of tackling the rotation period problem by means of what is known as the Doppler effect.

If an ambulance or a fire-engine approaches you while sounding its horn, you will notice that the note is high-pitched. As soon as the ambulance passes its closest point and begins to recede, the note of the horn deepens, because fewer sound-waves per second are entering your ear than would be the case for a stationary vehicle, and the wavelengths are effectively lengthened. There is an analogous effect with light; an approaching source of light will be "too blue", while a receding source will be "too red". The change in colour is too slight to be noticed in the ordinary way, but it involves a measurable shift in the positions of the dark lines, either to the red or to the blue end of the rainbow band. The effect was discovered over a century and a half ago by the Austrian scientist Christian Doppler – hence its name.

If Venus were in rapid rotation, one limb would be approaching

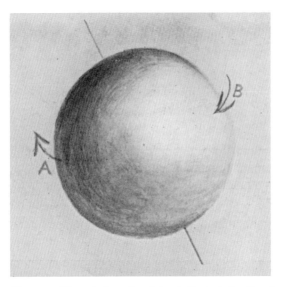

Diagram of the rotation of a globe, to illustrate the Doppler effect.

and the other receding. At point A, therefore, there would be a red shift; at B, a blue shift. Early results were negative, which did at least show that the rotation period must be long, certainly well over 24 hours. It was only in 1962 that the problem was solved, by the use of radar.

Radar contact with Venus was first achieved in 1961. The initial aim was to make a very accurate measurement of the distance of Venus, which would lead on to a determination of the length of the astronomical unit, or Earth-Sun distance; radar mapping of the surface would follow later. (In principle the method is straightforward enough. Radar pulses travel at the same speed as light; if you bounce a pulse off Venus and then record the "echo", the time taken for the pulse to travel from Earth to Venus and back again gives you the distance.) All the results indicated a very slow, retrograde (east-to-west) rotation. The true value is 243.01 Earth days, so that technically the Venus "day" is longer than the "year". The solar day on Venus is equal to 117 Earth-days, and if you could see the Sun through the clouds – which is never possible – the Sun would rise in the west and set in the east.

There was also another complication. At the Pic du Midi two

French observers, Royer and Guèrin, carried out a long study of Venus and recorded a persistent, Y-shaped feature centred on the equator. They even compiled a map. The Y-feature was usually seen, and from its movements Royer and Guèrin deduced a rotation period of only four days, retrograde. Surprisingly they were correct. The uppermost clouds really do have a four day period, even though the planet's globe spins so slowly. There is a strange shear effect in the atmosphere, unique in the Solar System.

The escape velocity of Venus is not much less than ours (6.3 miles per second, as against 7 miles per second for the Earth), and it would be logical to expect that the atmospheres of the two worlds would be similar in composition. Yet nothing could be further from the truth.

The first major advance was made in 1932 by the American astronomers Adams and Dunham, using the 100-inch reflector at Mount Wilson in California – then not only the largest telescope in the world, but in a class of its own. Adams and Dunham expected that the spectrum of Venus would show lines due to oxygen and nitrogen. Instead, they found that the main constituent of the atmosphere was the heavy, unbreathable gas carbon dioxide, and all later spectroscopic work led to the same conclusion. Carbon dioxide acts in the manner of a greenhouse, and shuts in the Sun's heat, so that clearly Venus had to be very hot indeed. On the other hand, direct measurements of the temperature of the upper cloud-layer, made with thermocouples, showed that the high atmosphere was decidedly chilly. The measured temperatures were of the order of -35°C (-30°F), and there was found to be little difference between the sunlit and the night hemispheres.

The atmosphere was found to be denser than ours; on the planet's surface the pressure is 90 times that of the Earth's air at sea-level, roughly equivalent to being under water on the sea floor at a depth of six miles. No evidence of water vapour was found in Venus' atmosphere, which was almost pure carbon dioxide.

Such was the state of our knowledge in 1961, when the first spacecraft to Venus was launched. It was indeed fragmentary. There was disagreement about the nature of the clouds, and also about the surface conditions; speculation was rife. It was even proposed that there might be broad oceans, and that the clouds could be of the same type as ours. If oceans had existed, they would have been fouled by the atmospheric carbon dioxide, and the result would have been seas of soda water – though, as I remember saying at the time, the chances of finding any whisky to mix with it seemed to be rather remote. Alternatively, Venus might be a raging dust-desert, without a scrap of moisture anywhere.

Nobody really knew. The rockets transformed the whole situation, and surprise followed surprise. But before dealing with space research developments, I must pause to discuss a few other topics – the Ashen Light, the phantom satellite, and transits and occultations.

7

The Ashen Light

When the Moon is a slender crescent in the evening sky, high enough to be seen against a reasonably dark background, the unlit "night" part can be seen shining dimly. This phenomenon is known to country folk as "the Old Moon in the Young Moon's arms", and has been known for centuries. Leonardo da Vinci was probably the first to give the correct explanation for it. It is due simply to light reflected from the Earth on to the Moon.

Terrestrial moonlight can be extremely strong, but earthlight on the Moon is even more brilliant, partly because the Earth will look larger in lunar skies but also because it is a much better reflector (its albedo is 0.37; that is to say, on average it reflects 37 per cent of the sunlight falling on it, while the average albedo of the Moon is slightly below 7 per cent). Telescopes show considerable detail on the earthlit part of the disk, and the phenomenon is visible every month up to the time when the Moon approaches half-phase.

A similar effect can be seen with Venus, though the cause is very different. Schröter recorded it, which is not surprising, but he was not the discoverer. The first record of the Ashen Light, as it is called, goes back to 1643, and is due to Johannes Riccioli, a Jesuit professor at Bologna. Riccioli's chief claim to fame is that he drew up a map of the Moon in 1651 on which he named the principal craters after people, usually astronomers, instead of keeping to the geographical analogies of his predecessor. Even though he can hardly be compared with men such as Huygens and the Cassinis, he was a competent observer, and there is no reason to suppose that he made a mistake with Venus, even though the Light was not

Earthshine on the Moon, photograph by Peter Wlasuk.

recorded again for some time.

William Derham, Canon of Windsor, saw the Light about 1714, and wrote:

'This sphaericity, or rotundity, is manifest in the Moon, yea in Venus too, in whose greatest falcations the dark part of their globes may be perceived, exhibiting themselves under the appearance of a dull and rusty colour.'

From a footnote to the third edition of Derham's book, published in 1719, it seems that he saw the Light frequently. (He was very much of an all-round scientist, and his investigations ranged from meteorology to biology and physics; he was also a medical practitioner. He was born in 1657, and died in 1735.)

There is little point in giving details of all later observations of the Ashen Light; suffice to say that there were plenty of them, and after about 1890 it was recorded by every serious observer of Venus – apart from Barnard, who was never able to see it. Normally it is visible only when Venus is a thin crescent. In 1895 Leo Brenner claimed to have seen it when Venus was more than half full, but, as we have noted, Brenner is emphatically not to be trusted.

The term "Ashen Light" should properly be restricted to the faint visibility of the night part of the disk, and should not be extended to cover reports of the night side appearing *darker* than the background, which must be due only to contrast effects. The French astronomer Camille Flammarion suggested that it might be due to the disk being seen against the subdued background glow of the Zodiacal Light, but this idea seems to be quite untenable.

As an example of a relatively modern description of the Ashen Light, it is worth quoting M. B. B. Heath, a very experienced observer who spent years studying Venus:

'When seen in daylight, the unilluminated portion of the disk was invariably noted as being darker than the surrounding sky, the darkness being frequently more prominent near the terminator and gradually shading off into invisibility at some point near the line

joining the cusps, but sometimes extending over the whole or nearly the whole disk. After sunset, at some rather indefinite time, the dark side has been noted as brighter than the outside sky, sometimes showing a dull red or brownish tint; usually this dim glow is seen over the whole unlit part of the planet. On two occasions it has been noted to be generally mottled, or of uneven brightness.'

Heath's observatory, at Kingsteignton in Devon, was equipped with a fine $10^1/_4$-inch reflector.

The first thing we have to decide is whether the Ashen Light is a real phenomenon, or whether it is merely a contrast effect, as has often been suggested. One problem is that it cannot be detected except against a darkish background, and this means that Venus must be low over the horizon. Also, the opportunities are restricted to the relatively brief period when Venus is a crescent. I have been following it ever since 1934, sometimes from my own observatory, where the main telescope is a 15-inch reflector, but sometimes from observatories with much larger instruments. On occasion, I have seen the Light so plainly that I cannot possibly dismiss it as illusory. For example, using my 15-inch reflector on the evening of 27 May 1980, just after sunset, the Light was so striking that it looked almost like the earthshine on the Moon, with the brightest part near the edge of the disk.

I worked out a way of testing the "contrast" theory. My plan was to hide the bright crescent, using a curved occulting bar in the eyepiece, and then see whether the Light remained visible. I found that as a rule it could still be seen shining "on its own", and to me this is conclusive; the Light is a genuine phenomenon.

The next step is to explain it, and speculation has not been lacking. A really weird proposal was made by a nineteenth-century German astronomer, Franz von Paula Gruithuisen, of the Observatory of Munich. Gruithuisen did some valuable work, but unfortunately his imagination was so vivid that even in his lifetime he was widely ridiculed. (For instance, he believed that he had discovered artificial buildings on the Moon.) Gruithuisen pointed out that the Ashen Light had been observed in 1759 (by Tobias

The Ashen Light sketched by Patrick Moore, 15-inch reflector x 300. The brightness of the Ashen Light is enhanced for reasons of clarity.

Mayer) and again in 1806 (by Schröter), an interval of 47 terrestrial or 76 Venus years, and went on to say:

'We assume that some Venus Alexander or Napoleon then attained universal power. If we estimate that the ordinary life of an inhabitant of Venus lasts 130 Venus years, which amounts to 80 Earth years, the reign of an Emperor of Venus might well last for 76 Venus years. The observed appearance is evidently the result of a general festival illumination in honour of the ascension of a new emperor to the throne of the planet.'

Later, Gruithuisen modified his theory. Instead of a Venus Coronation, he suggested that the Light might be due merely to the burning of large stretches of jungle to produce new farm land, and added that:

'Large migrations of people would be prevented, so that possible wars would be avoided by abolishing the reasons for them. Thus the race would be kept united.'

There are perhaps certain objections to these ideas of Gruithuisen's, and a saner theory was put forward by the German astrophysicist Hermann Vogel, who attributed the phenomenon to "a very extensive twilight". This sounds more plausible, but we are faced with the undoubted fact that the Light varies in intensity, which would hardly be the case if it were due to a twilight effect. The variations are often abrupt. For instance, I saw the Light well on 27 and 31 March 1953, but on 30 March I was unable to see a trace of it – and this was confirmed by independent observations in the United States, so that it was not due to any trick of the Sussex air. Needless to say, the observational conditions were standardized as much as possible with regard to altitude and background, while the seeing was virtually identical on the three occasions, and the telescope and eyepieces were the same.

Earthshine lights up the night side of the Moon; could the Ashen Light of Venus be due to illumination by some other body? The answer must be "no". Venus has no satellite, and therefore the only possible candidate is the Earth, which would admittedly be very brilliant as seen from Venus – or, more accurately, from above Venus' cloud-tops. But even elementary calculations are enough to show that this explanation is hopelessly inadequate. Earthlight will be far too weak to cause any perceptible glow.

Next there is a totally different theory, supported by William Herschel, who saw the Light many times around 1790, and also by Schröter. Venus might be largely ocean-covered, and the water would shine by phosphorescence. However, as soon as it was established that the surface is permanently cloud-covered, the "shining sea" idea had to be abandoned, and little can be said in favour of a proposal by Robert Barker, in 1954, that the Light is due to reflection from a layer of ice covering the entire planet!

Much more promising are the theories which involve aurorae in Venus' upper air. These go back to an original suggestion by P. de Heen as long ago as 1872, and at first sight there is nothing improbable about them. Many people – certainly those who live in high latitudes – are familiar with our own aurorae or polar lights (Aurora Borealis in the northern hemisphere, Aurora Australis in the southern), caused by streams of electrified particles sent out by the Sun which enter the upper atmosphere and make it glow.

The Earth is a powerful magnet, and the "magnetosphere" i.e. the region in space over which the magnetic field is dominant, is very extensive. The magnetic poles are not in the same positions as the geographical poles; they shift slowly, and at present the north magnetic pole is near north-west Greenland, while the south magnetic pole is in Antarctica. The magnetic field itself is due to movements in the Earth's liquid, iron-rich outer core, far below the surface of the globe. The Sun is sending out particles all the time, producing what is known as the solar wind; these particles cross the 93,000,000-mile gap between the Sun and the Earth, and reach the boundary of the magnetosphere, which acts in the manner of a barrier. The particles are deflected, and many of them are forced around the Earth's globe; some of them stream down into the upper air, and the result is an aurora.

Aurorae are high-level phenomena; the lower boundary lies at about 60 miles above sea-level, and the normal upper boundary is at around 185 miles, though in extreme cases aurorae may extend upward to as much as 600 miles. Because the solar particles are charged, they spiral downward toward the magnetic poles, so that aurorae are commonest in high latitudes. For example, they are almost always visible during nights in northern Norway and Alaska, and are common enough from Scotland, but brilliant displays over southern England are rare, and from Mediterranean countries aurorae are hardly ever seen. Displays cannot be predicted, but are commonest when the Sun is near the peak of its eleven-year cycle of activity, as it is during the first years of the present century.

Venus, more than twenty million miles closer to the Sun than we are, would be expected to have aurorae on a much grander scale. Some useful information has come from spectroscopic examination of the night side of the planet. Spectral lines similar to those recorded in terrestrial aurorae have been found, and from observations made with the 102-inch reflector at the Crimean Astrophysical Observatory, in 1955, the Russian astronomer Nikolai Kozyrev concluded that the brightness of Venus' night sky

Aurora Borealis seen from Tromsø, Norway. February 2000, Patrick Moore.

should be about fifty times greater than that of the night sky of Earth. On the other hand, Venus – unlike Mercury – seems to have no magnetic field at all.

I have attempted to correlate the visibility of the Ashen Light with events on the Sun – I reasoned that if auroral, the Light should be at its best when the Sun is active – but so far the results have been negative. The main difficulty about observations of this kind is, of course, that there are only limited periods when it is feasible to look for the Ashen Light. All the same, it seems that the problem gives plenty of scope to the amateur observer equipped with a moderate telescope and plenty of patience.

Another suggestion, made in 1999 by C. T. Russell and J. L. Phillips, is that the Light is due to consistent flashes of lightning. This too sounds plausible; after all, lightning has been detected in the atmospheres of both Jupiter and Saturn. We must also consider the permanent airglow in Venus' upper atmosphere, which is likely to be very strong. At least we are now confident that the Ashen Light is real, and that it is almost certainly electrical in nature.

8

Neith: the Phantom Satellite

Most of the planets in the Solar System have satellites. Some are really major, and two of them (Ganymede in Jupiter's system, and Titan in Saturn's) are larger than the planet Mercury. Only Mercury and Venus are unattended. With Mercury, there was an alarm on 27 March 1974, two days before the unmanned space-craft Mariner 10 was due to fly past; one instrument on Mariner reported ultra-violet emissions, indicating the presence of a satellite, but it then transpired that the object was an ordinary star, 31 Crateris. If Mercury had a satellite of appreciable size, it would certainly have been found by now.

The four brightest members of Jupiter's satellite family (Io, Europa, Ganymede and Callisto) were detected in 1610, and were observed by Galileo with his primitive telescope. In 1655 Huygens found Titan, the senior member of Satan's retinue. The satellite hunt was next taken up by G. D. Cassini, after his move from Italy to Paris. He was successful in finding four new Saturnian attendants, Iapetus, Rhea, Dione and Tethys. Then, in 1686, he made what was thought to be an even more important discovery. The following is an extract from his observational diary:

> '1686 August 18th, at 4.15 in the morning. Looking at Venus with a telescope of 34 feet focal length, I saw at a distance of $^3/_5$ of her diameter, eastward, a luminous appearance, of a shape not well defined, that seemed to have the same phase with Venus, which was then gibbous on the western side. The diameter of this object was nearly one quarter that of Venus. I observed it attentively for 15

minutes, and having left off looking at it for four or five minutes, I saw it no more, but daylight was by then well advanced. I had seen a like phenomenon, which resembled the phase of Venus, on 1672 January 25, from 6.52 in the morning, to 7.02, when the brightness of the twilight caused it to disappear. Venus was then horned, and this object, which was of diameter almost one quarter that of Venus, was of the same shape. It was distant from the southern horn of Venus a diameter of Venus on the western side. In these two observations, I was in doubt whether it was or was not a satellite of Venus, of such a consistence as not to be very well fitted to reflect the light of the Sun, and which in magnitude bore nearly the same proportion to Venus as the Moon does to the Earth, being at the same distance from the Sun and Earth as was Venus, the phases of which it resembled.'

It was then recalled that Fontana had seen something similar as far back as 15 November 1645. No further reports came in for some time, but near sunrise on 23 October 1740 the satellite was recorded by James Short, the well-known instrument-maker. His account of it is interesting enough to be reproduced in full:

'Directing a reflecting telescope, of 16.5 inches focus (with an apparatus to follow the diurnal motion) toward Venus, I perceived a small star pretty high upon her; upon which I took another telescope of the same focal distance, which magnified about 50 or 60 times, and which was fitted with a micrometer, in order to measure the distance from Venus; and found its distance to be about $10°2'0$. Finding Venus very distinct, and consequently the air very clear, I put on a magnifying power of 240 times, and, to my great surprise, I found this star put on the same phase with Venus. Its diameter seemed to be about a third, or somewhat less, of the diameter of Venus; the light was not so bright or vivid, but exceeding sharp and well defined. A line, passing through the centre of Venus and it, made an angle with the equator of about 18 or 20 degrees.

'I saw it for the space of an hour several times that morning; but the light of the Sun increasing, I lost it about a quarter of an hour

*after eight. I have looked for it every clear morning since, but never
had the good fortune to see it again.'*

Cassini, in his *Astronomy*, mentions another such observation:

*'I likewise observed two darkish spots upon the body of Venus, for the
air was exceeding clear and serene.'*

On 20 May 1759, at 8h 44m, Mayer reported the satellite:

*'I saw above Venus a little globe of inferior brightness, about 1 to
$1^1/_2$ diameters of Venus from herself ... The observation was made with
a Gregorian telescope of 30 inches focus. It continued for half an hour,
and the position of the little globe with regard to Venus remained the
same, although the direction of the telescope had been changed.'*

Two years later, further observations seemed to confirm the
satellite's real existence. A German astronomer, A. Scheuten,
reported that during the transit of 1761 he had detected a small
black spot following Venus across the Sun's disk, remaining visible
even when Venus itself had passed off the Sun. Next Montaigne of
Limoges, using a 9-inch telescope, produced a series of
observations which sounded most convincing.

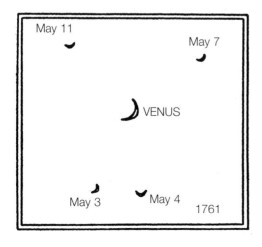

The position of Venus' satellite in
1761, according to Montaigne.

Montaigne first recorded the satellite on 3 May 1761, and described it as a little crescent-shaped body about 22 minutes of arc away from Venus. As usual, it showed the same phase as the planet itself, and had one-quarter the diameter of Venus. Montaigne repeated the observation several times during the night, and on 4, 7 and 11 May (the intervening nights were cloudy) he saw the companion again, differently placed but still showing the same phase. Montaigne, who had hitherto been decidedly sceptical about the existence of the satellite, was convinced. He added that he had taken every possible precaution against optical illusion, and that he had seen the companion even when Venus itself was put outside the field of view.

All this seemed definite enough. In a memoir read to the French Académie des Sciences, Baudouin announced:

'The year 1761 will be celebrated in astronomy, in consequence of the discovery that was made on 3 May of a satellite circling round Venus. We owe it to M. Montaigne, member of the Society of Limoges ... We learn that the new star has a diameter one-quarter that of Venus, is distant from Venus almost as far as the Moon from our Earth, has a period of 9 days 7 hours.'

In 1773 the German astronomer J. Lambert calculated an orbit which gave the mean distance from Venus as about 259,000 miles, with a period of 11 days 5 hours, an orbital inclination of 64 degrees, and an orbital eccentricity of 0.195. Frederick the Great of Prussia proposed to name the satellite "D'Alembert", in honour of his old friend Jean D'Alembert, but the prudent mathematician declined the honour with thanks.

Further observations were made on 3 and 4 March 1764 by Roedkiaer, from Copenhagen, on 10 and 11 March by Horrebow, also from Copenhagen, and on 28 and 29 March by Montbaron at Auxerre, who knew nothing about the Danish work. And henceforth, the satellite disappears from the observing books. Schröter could not find it, though he made a special search; neither could Herschel, and neither could Gruithuisen, who carried out a long series of observations. Satellites do not "softly and silently vanish away", like the hunter of the Snark, and there is no escape from the conclusion that the satellite of Venus never existed at all.

Maximilian Hell discussed the matter in 1766, and pronounced in favour of an optical illusion; later, von Ende wondered whether an asteroid might have been responsible, a view revived by Bertrand in 1875. At any rate, old myths die hard, and the ghost moon still had its supporters well into the second half of the nineteenth century. Admiral W. H. Smyth, author of the famous *Cycle of Celestial Objects*, believed in it, and claimed that "the satellite is perhaps extremely minute, while some part of its body may be less capable of reflecting light than others". This idea was developed in 1875 by F. Schorr, who went so far as to write a small book about it, *Der Venusmond*. Schorr revised Lambert's original period to 12d 4h 6m, and argued that the many failures to see the satellite were due to the fact that it varied in brightness, and was normally too faint to be visible. This theory sounded improbable even at the time, and in any case it was necessary to decide just what was meant by "the satellite". Whereas Cassini, Montaigne and others had described it as being a quarter the size of Venus itself, Roedkiaer and Horrebow, in Copenhagen, had seen it as a starlike point. Obviously something was wrong in the state of Denmark!

Another theory was announced in 1884 by J. Houzeau. Originally Houzeau had believed in the existence of the satellite, but then, following a full analysis, rejected it "first on account of the impossibility of properly representing the observed positions by an orbit described around Venus, and further because the mass of the planet derived from the least defective attempts would be seven times the real amount". On the other hand, Houzeau was reluctant to reject the actual observations, and he suggested that they could be explained not by a Venus satellite, but by a separate planet, moving in an orbit slightly outside that of Venus and with a period of 283 days. He even proposed a name for it: Neith, after an important goddess of Ancient Egypt.

The whole problem was more or less cleared up in 1887 by Paul

Stroobant of Brussels, who published an elaborate memoir in which he reprinted all the observations (thirty-three of them, made by fifteen different astronomers) and subjected them to a critical analysis. Some could be rejected outright, while others, such as Montaigne's, had to be put down to ghost reflections. Yet others could be attributed to faint stars. Horrebow, for instance, may have seen the fifth-magnitude start Theta Librae, while there is a chance that what Roedkiaer saw was the then unknown planet Uranus. Lambert's orbit, too, failed to survive the test of rigorous analysis, since it required the mass of Venus to be ten times greater than it actually is.

It would be easy to mistake a star (or even Uranus) for a satellite unless it were followed for long enough for its apparent motion to show it up in its true guise, and Venus is so brilliant that it is quite liable to give ghost images when seen against a dark background. (Per Wargentin, a famous Swedish astronomer of the eighteenth century, commented that one of his telescopes never failed to show companions to Venus or any other brilliant body.) Admittedly it is strange that observers so skilled as Cassini and Short fell into an elementary trap, but nobody is infallible.

Before leaving the subject, one more point must be mentioned. On 13 August 1892, Barnard, using the 36-inch Lick refractor, recorded a seventh-magnitude starlike object in the same field as Venus. The observation was made only half an hour before sunrise, and there is little chance of an optical ghost here – yet Barnard was able to measure the position, and it does not agree with any known star. However, Barnard had earlier made a special search for a satellite of Venus, and had satisfied himself that none existed. Joseph Ashbrook made the reasonable suggestion that Barnard had recorded a nova, or "new star", which by bad luck was not seen by anybody else.

Though Venus still presents us with many problems, this one at least may be regarded as finally solved. Had there been a satellite of reasonable size, it could not have escaped detection in the Space Age. "Neith" does not exist, and has never done so. Venus, like Mercury, is a lonely wanderer in space.

9

Transits and Occultations

As the Earth's distance from the Sun is greater than that of Venus, there must be occasions upon which the three bodies move into a direct line, with Venus in the middle. This, of course, can occur only at inferior conjunction. When the alignment is perfect, Venus can be seen with the naked eye as a black spot silhouetted against the solar disk. This is known as a transit of Venus.

Were the orbits of the two planets in the same plane, a transit would happen at every inferior conjunction, but unfortunately this is not the case. The orbit of Venus is inclined to ours at the small but significant angle of 3.4 degrees, and this is enough to ensure that transits are rare. At the present epoch, they are seen in pairs, the components of the pair being separated by eight years, after which no more transits occur for over a century. Thus there were transits in 1631, 1639, 1761, 1769, 1874 and 1882; the next will be on 7 June 2004, 5 June 2012, 10 December 2117, 8 December 2125, 11 June 2247, 8 June 2255, 12 December 2360 and 10 December 2368. Go forward to the year 3818, and there will be a transit on Christmas Day – if Christmas is still celebrated as far ahead as that!

At other inferior conjunctions, the planet passes above or below the Sun in the sky, and escapes transit, though telescopically it can be followed almost without a break. In 1950, for instance, photographs taken with the 16-inch telescope on Mule Peak, New Mexico, showed Venus when only 7.5 degrees from the centre of the Sun.

Represent the orbits of the Earth and Venus by two hoops, crossing each other at an angle of 3.4 degrees. The points where

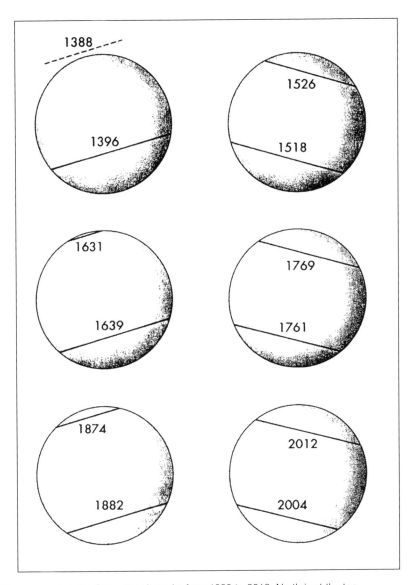

Venus crossing the Sun – transit tracks from 1388 to 2012. North is at the top.

the two hoops cross are known as the nodes. Half of Venus' orbit is "above" ours and the other half "below" (technically one ought to say "north" and "south", because there is no fixed reference in space, and so no true "up" or "down"). Venus takes 224.7 Earth-days to go once round the Sun. For a transit to occur, Venus must be at, or very near, a node at the time of inferior conjunction. The plane of Venus' orbit passes through that of the Earth twice a year, on 6 June and 7 December. This means that a transit can occur only on, or very near, one or the other of these dates. There is a certain amount of leeway, but not much, because the orbit of Venus is nearly circular; the eccentricity is only 0.007, whereas that of the Earth's orbit is 0.017.

It so happens that eight Earth years are very nearly equal to thirteen Venus years, so that after eight years the two planets and the Sun are in nearly the same relative positions (the vital word here is "nearly"). A transit may therefore be followed by another transit eight years later, as happened in 1874 and 1882. But in 1889, when the relative positions were again nearly the same, the alignment was no longer exact enough for a transit to occur, and observers had to wait patiently for the next pair, due in 2004 and 2012.

To sum up: at the present epoch transits occur in pairs, each pair being separated from the next by over a century. At least a transit of Venus is a leisurely affair, and there is no need for frantic last-minute preparations as there is for a total eclipse of the Sun. For example, the transit of 8 June 2004 will begin at 05.13.27 GMT and end at 11.25.57 a total duration of over six hours. (If clouds cover the sky, you will even have time to hire an aircraft!)

Obviously, the only large members of the Solar System to be seen in transit are Mercury and Venus. Numerous small asteroids can also do so, but are too tiny to be detected against the face of the Sun, and even Mercury is too small to be seen with the naked eye during transit. Not so with Venus, which is conspicuous as a black disk – though at the present time (2002) there can be no living person who can remember a transit of Venus.

The first transit predictions were made by Johannes Kepler, the great German mathematician who was the first to work out the ways in which the planets move round the Sun, and who drew up the Laws of Planetary Motion upon which all later work has been based. In 1627 he finished what proved to be his last work, a set of new and more accurate tables of planetary motions named the

"Rudolphine Tables" in honour of his old benefactor Rudolph II. From these, he was able to calculate that both Mercury and Venus would transit the Sun during the year 1631, Mercury on 7 November and Venus on 6 December. By that time Kepler was dead, but the transit of Mercury was successfully observed by the French astronomer Pierre Gassendi.

Encouraged by this success, Gassendi naturally expected to be equally fortunate with Venus, which is not only much closer to us than Mercury but is also much larger. He left nothing to chance. Fearful that Kepler's prediction might be in error, he began watching the Sun on 4 December (between breaks in the clouds) and continued observing all through 6 and 7 December. He saw nothing. The reason is now known; the transit did indeed occur, but took place during the northern night of 6-7 December, when the Sun was below the horizon in France.

Kepler had predicted no more transits of Venus before 1761; he

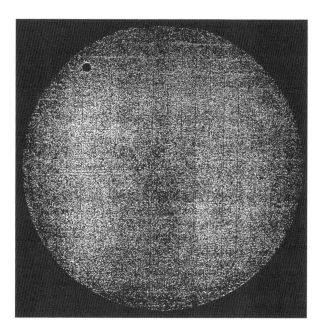

The 1874 transit of Venus, photographed by the pioneering astrophotographer Jules Janssen. Although not clear by modern standards, the image does show the apparent size of Venus relative to the Sun.

Carr House, Hoole, scene of the first observation of Venus in transit.

had missed the fact that they occur in pairs, separated by eight years. And this is where Jeremiah Horrocks comes into the story.

Horrocks was born at Toxteth, near Liverpool, in 1619. Not much is known about his early years, but clearly he showed great ability, and by 1632 he had entered Emmanuel College, Cambridge. It was here that he became seriously interested in astronomy, and he made some notable friends, either by personal contact or by correspondence; among them were William Crabtree, of whom more anon, and William Gascoigne, inventor of the micrometer, who was later to "lose his life during the Battle of Marston Moor between Charles I's Cavaliers and Oliver Cromwell's Roundheads". Horrocks was emerging as a mathematician of the first rank, and late in October 1639 he made a careful examination of Kepler's calculations for transits of Venus. He found something very interesting: there might be a transit in December of that year. He was by no means certain – there was a distinct chance that Venus might pass clear of the Sun – but it was worth following up. Time was short, and all Horrocks could do was to send messages to Crabtree, who was living in Salford, and his brother Jonas Horrocks, in Toxteth. Postal services in those days were very slow, and it is unlikely that anyone else had the slightest inkling that a transit might occur.

Details of Horrocks' career around this time are sparse. It is often said that he had entered the Church, and had become curate at a church in Hoole, a village fifteen miles from Liverpool, so that in most books he is referred to as "the Rev". In fact, according to

Crabtree observing the transit of 1640 – as painted 200 years later by Ford Madox Brown. The drawing is certainly inaccurate!

Allan Chapman, Britain's leading astronomical historian, there is no evidence that he was ever ordained, and a good deal of indirect evidence that he was not. In 1639 he was only twenty years old, too young to have been a deacon and much too young to be a full priest. He had the benefit of a Cambridge education, and may have supported himself by teaching. At any rate, at the time of the transit he was certainly in Hoole, and associated in some way with the local church, but by the summer of 1640 he was back home in Toxteth with his family, who had no clerical connections.

Horrocks made what preparations he could, and there is no

doubt that he observed the transit from an upstairs window in Carr House, close to the church. His telescope was modest by modern standards; apparently it was a Galilean refractor, but its precise aperture is not known, and unfortunately Horrocks did not tell us. Sensibly, he used the telescope to project the Sun's image on to a graduated screen fixed behind the eyepiece, as Gassendi had done when following the transit of Mercury eight years earlier; the image on Horrocks' screen was six inches across. He claimed that he was able to see "the smallest spots on the Sun", and during the transit he noted that Venus was much blacker than any sunspot could be.

It was natural for him to be cautious, and he began observing well ahead of time. He watched the Sun for much of the day before the predicted transit, and saw nothing apart from a few spots. On the morning of "transit day" itself, clouds covered the sky, and moreover Horrocks was called away by urgent duties – precisely what these duties were, we do not know. But by 3.15pm he was back at the telescope, and then:

> *'At this time, an opening in the clouds, which rendered the Sun distinctly visible, seemed as if Divine Providence encouraged my aspirations; when, O most gratifying spectacle! the object of so many earnest wishes, I perceived a new spot of unusual magnitude and of a perfectly round form, that had just wholly entered upon the left limb of the Sun, so that the margin of the Sun and spot coincided with each other, forming the angle of contact.'*

He followed the planet until sunset, an hour and a half later, and was able to make some useful measurements; in particular he checked on the apparent diameter of Venus, and found it to be about one minute of arc, smaller than most astronomers had expected. (As seen from Earth, the maximum apparent diameter is in fact just over 1 minute 5 seconds of arc.) Crabtree had bad luck with the weather, but did manage to see Venus just before sunset when the clouds broke up for a few minutes. (In Manchester City Hall there is a painting of Crabtree observing the transit. Alas, it does not seem to be authentic, but at least it is a magnificent picture!) Jonas Horrocks seems to have missed the transit altogether; in any case, he left no record of it.

Jeremiah Horrocks' prediction was a brilliant piece of work, and it was by no means his only achievement.* He was clearly destined for fame, but, sadly, died in 1641 at the early age of twenty-two.

---

* In 1960 – when television was still "live", and black-and-white – we decided to present a programme about transits of Venus in my *Sky at Night* series. We went to Carr House, which was then a doll museum, and set up our cameras. As introductory music, we used *At the Castle Gate*, from the suite *Pelleas and Melisande*, by Sibelius. Half an hour before the programme was due to be transmitted, we realized that the Sibelius record had been left behind. Crisis! We searched frantically for a piano; there wasn't one. We did find an ancient harpsichord. I had never played a harpsichord, and I had never played *At the Castle Gate*. I had about ten minutes to rehearse – and I bravely played us "on" and "off" the air. What Jean Sibelius would have thought, I do not know.

Crabtree survived him by only three years. He was older than Horrocks - he was born in 1610 – and became a successful merchant; he was a self-taught astronomer, and made some useful measurements. It has been said that he was killed during the Battle of Naseby, between the Cavaliers and the Roundheads, but I have not been able to confirm this.

At least astronomers were prepared for the next transit, that of 1761, and during the interim years there was an important development. Edmond Halley, following up an earlier suggestion by the Scottish mathematician James Gregory, found that it would be possible to use transits of Venus to measure the length of the astronomical unit, i.e. the distance between the Earth and the Sun. In theory, transits of Mercury could be used in the same way, because Halley's method was essentially geometrical, but Mercury's disk was too small for precise measurement, as Halley realized.

At that time the distance of the Sun was not known with real accuracy. The best estimate had been made by Cassini, in 1672; he gave a value of 86,000,000 miles, while Flamsteed, the Astronomer Royal, preferred 81,000,000 miles. Kepler's Laws made it possible to draw up a complete scale map of the Solar System by following the movements of the various bodies, but what was needed was one absolute value. If this could be obtained, everything else would fall neatly into place.

In practice, what has to be done is to time the transit from entry on to the Sun's disk (ingress) to exit (egress). The timing must be done with synchronized clocks which are in exact agreement. Each observer records the times of ingress and egress; the observed track across the Sun's disk can then be found, and that is all that we need.

The method is perfectly sound – in theory. Halley pointed out that the observers should be separated as widely as possible; he suggested North Norway and the East Indies. The main problem was in dispatching expeditions to very widely separated sites, which was much more difficult then than it would be now. There was a different approach, due to the French astronomer Joseph De

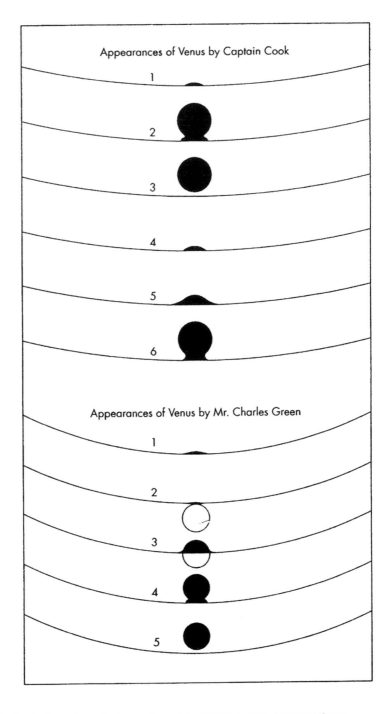

The Black Drop, shown in observations of the 1769 transit by Cook and Green.

L'Isle, which depended only upon the exact moment of ingress, but this method required a very accurate knowledge of the longitudes of the observing sites, and this sort of information was not readily available in the eighteenth century.

Halley could not hope to see the next transit – he died in 1742 – but his advice was enthusiastically followed. Observations were made by almost two hundred astronomers from over 120 sites, despite the fact that England and France were, as usual, at war with each other.

There were two British expeditions. One, led by the Astronomer Royal (Nevil Maskelyne) went to the island of St Helena, but at the critical moment the sky was cloudy. The second party, led by Dr John Winthrop, went to St John's in Newfoundland, and were able to make excellent observations. The French had mixed fortunes. One party, led by the Abbé Jean Chappe d'Auteroche, was completely successful from Tobolsk in Siberia; the other main expedition, led by Alexander Pingré, went to the Indian Ocean island of Rodriguez, but saw ingress only.

Unfortunately, a cruel trick of Nature affected all the results. When Venus passes on to the Sun, it seems to draw a strip of blackness after it – and when this strip disappears, the transit has already begun, so that ingress cannot be timed accurately. The "Black Drop", due mainly to Venus' atmosphere, meant that the final results from the 1761 transit were disappointing, and gave different values for the astronomical unit, ranging between 96,162,840 miles and only 77,846,000 miles.

No account of the 1761 transit would be complete without relating the sad story of the French astronomer Legentil – or, to give him his full name, Guillaume Joseph Hyacinthe Jean Baptiste Le Gentìl de la Galasière, whose expedition must have been the unluckiest in the whole history of science. In March 1760 he set sail for Pondicherry in India, where conditions for the transit were expected to be very good. The first part of his journey, to the Isle de France, was reasonably uneventful apart from a near-encounter with British warships off the Cape of Good Hope. Then, to his consternation, he learned that Pondicherry had fallen to the

Tablet set up at Tahiti, where the transit of Venus was observed.

British, and all Legentil could do was to head back to the Isle de France. He had a perfect view of the transit – but from the deck of a ship, and could make no useful observations of any kind.

What next? Rather than return home empty-handed, he elected to remain in the East until 1769, and observe the second transit instead. He was not idle, and in fact he made a great many varied and valuable scientific observations, but the transit was always uppermost in his mind, and he gave careful consideration to the choice of site. He finally decided upon Manila, in the Philippines, and arrived there in August 1766, which certainly gave him plenty of time to prepare.

Then, in July 1767, Legentil received a letter from France. The French authorities were very anxious for him to go back to Pondicherry rather than stay in the Philippines. Finally Legentil agreed, possibly because he had run foul of the corrupt Governor of Manila, and by March 1768 he was back in Pondicherry. This time he had full co-operation from the British, who even loaned him an excellent 3-foot telescope, and he set up his equipment.

On 2 June 1769, the day before the transit, the weather was perfect. Then the clouds rolled in; Legentil had to wait under an overcast sky, praying for a miracle which did not happen. Half an hour after the end of the transit, the sky was crystal clear again ... and then it transpired that at Manila there had been no cloud cover at all.

One can well imagine Legentil's feelings. It was rather too long to wait for the next transit (that of 1874), so at last he decided to go home. After two minor shipwrecks he reached Paris in October 1771, to find that he had been presumed dead and that his heirs were preparing to distribute his property; he took legal action to reclaim it at something later than the eleventh hour. It seems likely that by then his enthusiasm for transits of Venus was somewhat dampened.

In was in 1761 that Mikhail Lomonosov made the observations which led him to infer the presence of an atmosphere surrounding Venus, and at the 1769 transit David Rittenhouse, in America, saw that the edge of that part of the planet which was off the solar disk seemed to be illuminated, so that the whole outline of Venus was visible; this could only be due to an atmosphere.

Despite the menace of the Black Drop, astronomers were determined to make the most of the transit of 1769. Nevil Maskelyne was particularly enthusiastic, and at a special Royal Society meeting on 17 November 1767 he was one of four Fellows making firm recommendations (the others were John Bevis, James Short and James Ferguson). Eventually, parties were sent to Hudson's Bay (William Wales and James Dymond), North Cape (William Bayly), Hammerfest (Jeremiah Dixon) and north-west Ireland (Mason). Finally there was the South Seas expedition, which brings us to the story of Captain James Cook.

A ship was fitted out expressly for the purpose, with the full agreement of King George III, who was an astronomical enthusiast and who actually watched the 1769 transit from the newly-founded observatory at Kew. The chosen observer was Alexander Dalrymple, but when he was refused command of the ship he withdrew in a huff, and the Admiralty called in Cook instead. The senior astronomer was Charles Green, who had been an assistant at Greenwich Observatory and was well known to Maskelyne. Also on board was a distinguished Swedish scientist, Daniel Solander, who was primarily a botanist but who had also an excellent knowledge of astronomy. After some deliberation the Admiralty decided that the island of Tahiti would be the best site, and on 26 August 1768 the *Endeavour* sailed from Plymouth. By 10 April 1769 the ship was standing off Tahiti, or, to give it the natives' own name, Otaheite. (Europeans had first heard of it in 1607, when it had been discovered by Pedro Quiros; Captain Samuel Wallis rediscovered it in 1767 and named it King George's Island, which is what Cook called it.)

A temporary observatory was set up, and there were no mishaps, apart from the fact that an essential quadrant was stolen by one of the local inhabitants, and was recovered only with some difficulty. On the day of the transit the weather was perfect, and excellent observations were made by Cook, Green and Solander. Cook's own account read as follows:

> 'The first appearance of Venus on the Sun was certainly only the penumbra, and the contact of the limbs did not happen till several seconds after ... it appeared to be very difficult to judge precisely of the times that the internal contacts of the body of Venus happened, by reason of the darkness of the penumbra at the Sun's limb, it being there nearly, if not quite, as dark as the planet. At this time a faint light, much weaker than the rest of the penumbra, appeared to converge toward the point of contact, but did not quite reach it. This was seen by myself and two other observers, and was of great assistance to us in judging of the time of internal contacts of the dark body of Venus, with the Sun's limb ... I judged that the penumbra was in contact with the Sun's limb 10″ sooner than the time set down above; in like manner at the egress the thread of light was wholly broke by the penumbra ... The breadth of the penumbra appeared to me to be nearly equal to $\frac{1}{2}$ of Venus' semi-diameter.'

Cook gave the apparent diameter as 56 seconds of arc, which was very near the truth.

The *Endeavour* left Tahiti on 13 July, and continued on the voyage of exploration which is part of our history. The real tragedy is that when the ship was near Java, Charles Green died.

The main French expedition also had a tragic end. It was led by Chappe, who had been so efficient in 1761. It was sent to San Jose del Cabo in Baja California. (Originally the British had cast their eyes upon Baja California, but the area was then controlled by Spain, and relations between the British and Spanish Governments at that time were decidedly frosty.) Chappe and his team had no trouble in setting up a makeshift observatory, but there was one terrible threat; disease, which seems to have been a particularly virulent form of typhus. At least one-third of the local population had died, and it would have been prudent for the astronomers to leave, but instead they decided to stay and carry out their programme. The observations were successful – but then the disease stuck. Chappe and his main colleagues died, and only two members of the expedition survived to return home.

One other story about the 1769 transit is worth retelling. The Hungarian astronomer Father Maximilian Hell, Director of the Vienna Observatory, observed the transit from Vardø, an Arctic island off the coast of Lapland. His programme was completely successful, and the results were published in full in 1772. Much later, Karl von Littrow, who became Director of the Observatory, examined Hell's diaries and came to the startling conclusion that they had been fudged; some entries had been crossed out and rewritten in ink of a different colour. Von Littrow published his accusations, and Hell's reputation was ruined.

Then, in 1883, the diaries were re-examined by Simon Newcomb, one of America's most eminent astronomers. Newcomb proved

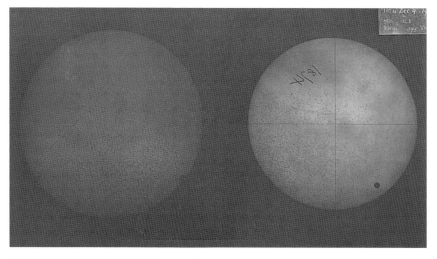

Two photographs of the 1874 transit of Venus.

that Hell's observations were perfectly genuine; the alterations had been made at the time, and different ink had had to be used in the Arctic chill. Von Littrow had jumped to his conclusions for a curious reason: he was colour-blind, and could not even tell a red star from a white one. Hell's reputation was vindicated. In a letter to Newcomb, John Hagen of Georgetown Observatory – like Hell, a Jesuit – wrote:

'It was fitting that this act of justice should be reserved to an American astronomer who stands aloof from the petty quarrels of the old world.'

The 1769 transit was observed from ten stations, and there were 150 sets of measurements. Yet mainly because of the Black Drop, the results were still disappointing. They were carefully analysed by the Finnish mathematician Anders Lexell, who gave a value for the astronomical unit of 95,000,000 miles. In 1824 Johann Franz Encke, of the Berlin Observatory, made a new analysis, which gave almost the same result. The modern value is 92,957,000 miles, so that the results by Lexell and Encke gave a figure which was considerably too high.

There the matter rested for over a century, but there was still optimism about the transits of 1874 and 1882, and a huge international programme was planned; by then, of course, photography had become sufficiently good to be of real help. British parties went to Egypt, India, New Zealand, Australia, Mauritius, Pekin and the Antarctic island of Kerguelen, which was (and still is) one of the most inaccessible places in the world. Conditions at many of the observing sites were good, but yet again the results were discordant. Final attempts were made at the transit of 1882, and British parties went to sites ranging from Queensland and South Africa to Bermuda: the observatory at Natal was set up specially for the transit, and observations were made by Edmond Nevil (better known as Neison) whose main claim to fame is that he produced an excellent map of the Moon. Sadly, the results were no better than before, and despite all the care and effort it had to be admitted that transits of Venus simply were not able to yield an accurate value for the length of the astronomical unit.

This means that the next transits, those of 8 June 2004 and 6 June 2012, will be regarded more as interesting spectacles than as major astronomical events. Clouds permitting, the 2004 transit should be well seen from the British Isles; from Greenwich ingress will be at just after a quarter past five in the morning (GMT) and Venus will not pass off the solar disk until almost half-past eight.* Sunrise at Greenwich on that day is at 03h 45m, an hour and a half before ingress; Venus will cross the southern hemisphere of the Sun. The 2012 transit will be seen from the Far East; Britain will have a tantalising glimpse of Venus just about to egress as the Sun rises. Japan will be an ideal observing site. After that we must wait for 11 December 2117 and 8 December 2125.

Transits of Venus could of course be observed from other planets – if we could get there! From Mars, Venus will pass in transit on 20 August 2030, 18 June 2032, 5 November 2059 and 17 June 2064 (a Martian observer would, incidentally, see a transit of the Earth on 10 November 2084). On 16 July 1910 a transit of Venus would have been seen from an observer on Saturn, lasting for over eight hours, but to a "Saturnian" the apparent diameter of Venus would be less than two seconds of arc.

Ironically, Venus did in the end provide a means of measuring the astronomical unit – but not by the transit method. A radar pulse was bounced off the planet, and the echo was received. We know the speed of a radar pulse, and therefore the time taken for the there-and-back journey gave the distance.

As the Moon moves across the sky, it sometimes passes in front of a star and hides or occults it. When this happens, the star remains visible right up to the lunar limb, and then snaps out abruptly, since the tenuous lunar atmosphere is far too thin to have any appreciable power of dimming or refraction. (The entire

---

* For full details, see *Transit: When Planets Cross The Sun*, by Michael Maunder and myself (Springer Verlag, 2000)

Occultation of Venus,
Japanese photo 1912.

weight of the Moon's atmosphere, if it could be measured under terrestrial conditions, would be no more than about 30 tons.) A star is to all intents and purposes a point source, but a planet shows a disk, and naturally an occultation by the Moon is a more gradual process; it takes some seconds for the disk to be covered by the onrushing limb of the Moon. Venus is occulted now and then, and these events are interesting to watch, if not of any actual value.

Very rarely, one planet may be occulted by another. This happened on 3 October 1590, when an occultation of Mars by Venus was seen by Michael Möstlin, Professor of Mathematics at Heidelberg; and again on 17 May 1737, when Mercury was occulted by Venus. On 21 July 1859 Venus and Jupiter were so close together that they could not be separated with the naked eye, though there was no actual occultation.

When Venus occults a star, useful observations can be made. Before being covered, the star shines for a few moments through the Venus atmosphere, and the amount by which it is dimmed or reddened gives a clue as to the depth of the gaseous mantle surrounding Venus. On 26 July 1910, the star Eta Geminorum (Propus) was occulted, and observations were made at the Flammarion Observatory at Juvisy, near Paris, by E. M. Antoniadi, F. Baldet and F. Quénisset, using telescopes of up to 9 inches aperture. Their report read as follows:

'The emersion occurred under favourable conditions, so that we were able to confirm, independently and very clearly, that Eta Geminorum [then of magnitude 3.5] * *did not reappear suddenly, as with lunar occultations. In fact, there was at first a barely perceptible luminescence; then the very faint star seemed to detach itself from the dark edge of the planet. It rapidly increased in brilliancy, and in 1.5 to 2 seconds after its first appearance it had regained its initial brightness. Besides this increase in luminosity at the moment of emersion, we notice that Eta Geminorum continued to gain slowly and slightly in intensity according to its distance from Venus … There was no appreciable change in the colour of the star … The hypothesis which appears to us the most probable for explaining the variation in brightness is that the light of the star was absorbed in traversing the atmosphere of Venus. In our observation, this variation, which lasted from 1.5 to 2 seconds, corresponds to a motion of the planet from 0".8 to 1".1. We hence deduce the height of the atmosphere of Venus which produced this absorption to be from 50 to 70 miles.'*

This sort of behaviour was confirmed in 1918, when the star 7 Aquarii was occulted, and on 19 March 1948, when 36 Arietis was the occulted star. On 7 July 1959 Venus occulted the bright star Regulus (Alpha Leonis); immersion was at 14:28 GMT, broad daylight, and I was able to observe it from Selsey, using a 12-inch reflector. The fading before occultation was very pronounced, and gave a value for the height of Venus' atmosphere that proved to be very near the truth. The next Venus occultation of a really bright star (Regulus, again) will be postponed until 1 October 2044.

It cannot be said that occultations are now of real importance, as they once were, but they are always interesting to watch, and it is fascinating to see a star flicker and fade before being finally hidden as the solid body of Venus sweeps over it.

---

* Eta Geminorum is a red semi-regular variable star of spectral type M; the range is from magnitude 3.2 to 3.9, and there is a very rough period of about 233 days.

10

Rockets to Venus

Up to now, I have been describing how astronomers set about unravelling the secrets of "the planet of mystery". With the opening of the Space Age, with the launch of Russia's Sputnik 1 in October 1957, the whole situation changed. And in retrospect, it is interesting to look back at the state of our knowledge (or lack of it) less than fifty years ago.

It had been established that the main constituent of the Venus atmosphere was carbon dioxide, but there was no positive information about the state of affairs beneath the cloud-tops. Presumably the surface temperature was high, but the various estimates did not agree at all well. The length of the rotation period was most uncertain; G. P. Kuiper's estimate of "a few weeks" was regarded as the most likely value, but there were still some astronomers who preferred a period in the region of 24 hours, and in France the 224.7-day synchronous period remained popular. Efforts to map the surface features had led to no really useful results. As for the surface itself – well, it could be either an arid, fiercely hot dust-bowl, or else covered with water. If there were seas, there seemed to be no valid reason why primitive life-forms should not have appeared, just as they did in the warm pre-Cambrian seas of Earth. Venus, it was said, might be a world where life was just starting to develop.

Then, on 26 August 1962, Mariner 2 was sent on its way. The new era in Man's exploration of Venus had begun.

Actually, Mariner 2 was not the first Venus probe, as there had been several previous attempts, all of them unsuccessful. The

This view of Venus was taken by Mariner 10's television cameras on February 6, 1974, one day after Mariner 10 flew past Venus on the way to its first encounter with Mercury. The images were taken in invisible ultraviolet light and show features in the planet's clouds that had not been observed before. This photo was made by first computer enhancing several television frames at JPL and then mosaicking and re-touching them at the U.S. Geological Survey's Division of Astrogeology at Flagstaff, Arizona. The blue colour was applied to the image and does not represent the actual colour of Venus' clouds.

Artist's impression of Mariner 2, the first spacecraft to successfully send back information from Venus in December 1962.

Russians had taken the lead; after all, they had launched the first satellite, sent the first rockets past the Moon, and were about to send the first man into space (Yuri Gagarin). They began their interplanetary programme with Venera 1, in February 1961, and the probe was put into the correct orbit, but before long contact with it was lost. The first American attempt, with Mariner 1 in July 1962, also failed; the spacecraft simply plunged into the sea, apparently because someone had forgotten to feed a minus sign into the computer (a slight mistake which cost approximately £4,280,000). So Mariner 2 became the first spacecraft to send back data direct from Venus.

Sending a probe to another planet is not so straightforward as might be expected. It would be convenient to wait until the Earth and the target planet were at their closest together (around 24,000,000 miles in the case of Venus) and then fire a rocket straight across the gap, but there are any number of reasons why this cannot be done. For instance, it would involve using more propellant than any rocket could possibly carry. The procedure is to make use of the Sun's gravitational pull, so that the probe can "coast" along, unpowered, for most of the journey.

First, the probe is put into a "parking" orbit round the Earth, using a launch vehicle of the usual type. Then, at the correct moment, a further burst of power makes the probe leave the region of the Earth for good, and enter an independent orbit. This is so arranged that the probe moves away in the direction opposite to the Earth's movement round the Sun, so that, broadly speaking, the probe will be travelling round the Sun at a rate rather less than the Earth's own 66,000 mph ($18^1/_2$ miles per second). This means that it cannot follow the Earth round; it

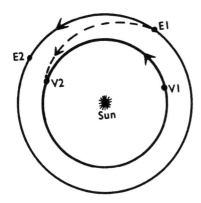

A typical transfer orbit: A probe is launched with the Earth at E1 and Venus at V1. The probe swings inward, and meets Venus at V2. Meanwhile the Earth has moved on to E2.

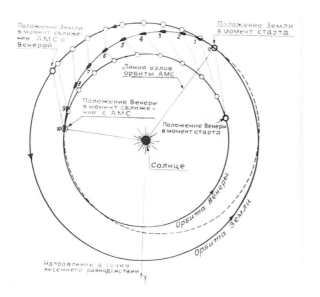

Положение Земли
в момент сближе-
ния АМС с
Венерой

Положение Земли
в момент старта

Линия узлов
орбиты АМС

Положение Венеры
в момент сближе-
ния с АМС

Положение Венеры
в момент старта

Солнце

Орбита Венеры

Орбита Земли

Направление в точку
весеннего равноденствия

Russian chart showing the planned orbit of the failed 1961 Venera 1 probe.

must start to swing inward, and if all goes well it will reach the orbit of Venus and meet the planet at a prearranged point. The diagram will show what is meant – I have likened it to the cosmic equivalent of clay-pigeon shooting. If there is no collision, the probe will begin to swing outward again, and will continue to move round the Sun in an elliptical orbit.

A path of this kind, taking a spacecraft from the orbit of one planet to that of another, is called a transfer orbit. The principle had been worked out long before rockets had been developed, and some writers, notably Walter Hohmann in Germany, had produced very detailed plans.

A list of the Venus probes launched between 1961 and 2001 is given in the Appendix. I have omitted some dubious cases; until fairly recently the Russians were very secretive, and they may well have made several unsuccessful attempts in the 1960s and 1970s. Mercifully the situation is different today, and space research has become genuinely international.

Venera 1, in February 1961, blazed the trail. It was a complex affair, and initially it seemed to be very promising; the launch was perfect, and Venera sped on its way. Unfortunately, at this time the Russians were having communication problems, and on 19 February, at a distance of over a million miles from Earth, Venera 1 "went silent". Contact was never regained, and so what happened to the probe we do not know, though there is every chance that it went within 65,000 miles of its target in the following May, and then continued in orbit round the Sun. The best that could be said of it was that it was encouraging. There had been no intention of hitting Venus; the idea was to obtain information from close range, and send back the data to be decoded at leisure.

In the following year the Mariner programme was started from Cape Canaveral. Following the Mariner 1 fiasco, Mariner 2 was put into orbit, on 27 August, with very different results.

As usual, the lower stage of the launcher was an Atlas rocket which rose vertically from Cape Canaveral and then moved off in the general direction of South Africa. The second stage, an Agena rocket, then took over, and the Agena-Mariner combination was put into a parking orbit at a height of 115 miles, moving at the regulation 18,000 mph. As it reached the African

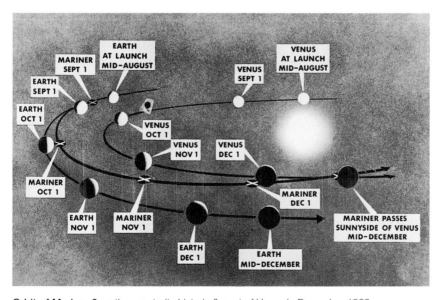

Orbit of Mariner 2 on the way to its historic flypast of Venus in December 1962.

coast, the Agena fired once more, and the total velocity rose to 25,503 mph, which is about 850 mph more than the escape velocity for that height.

Remember that for a Venus mission, the direction of "escape" has to be in the direction opposite to that in which the Earth is moving round the Sun. It was so on this occasion. As the Mariner moved away it was slowed down by the Earth's gravitational pull, until at a distance of 600,000 miles the velocity relative to the Earth had fallen to only 6,874 mph. In other words, Mariner was moving round the Sun at a velocity of 6,874 mph, less than that of the Earth, and it started to swing inward toward the Sun, picking up speed as it went. Meanwhile, the Agena rocket had been separated from Mariner, rotated through a wide angle and fired again, so that it was put in to an entirely different orbit. Nobody knows what happened to it, and nobody cares; its work was done, and on the journey to Venus it would simply have been a nuisance. Henceforth, Mariner 2 was on its own.

This was only the beginning. Mariner was more or less in the right orbit, but it had to be kept there, and this was achieved by making it "lock" on to the Earth and the Sun. The aim was to send the probe past Venus at a range of about 9,000 miles; when it reached Venus' orbit it was expected to have a velocity of 91,000 mph relative to the Sun, as against the 78,000 mph of Venus itself. This would make for a reasonably prolonged encounter.

It would have been over-optimistic to hope for complete success at what was to all intents and purposes a first attempt (Mariner 1 can hardly be counted, as its flight lasted for only a few minutes). A mid-course correction was made on 4 September, but even so the minimum distance from Venus was 21,594 miles instead of less than 10,000. Fortunately, the instruments had been designed to operate over a considerable range, and the error was not ruinous.

The date of closest approach was 14 December. Contact was maintained without difficulty, and a great deal of information was sent back. It took some time to analyse and interpret the data, but by February 1963 the astronomers were able to give some positive results. To say that some of these findings were unexpected is to put it mildly.

First, and perhaps most important of all, Venus was super-tropical, with a surface temperature of something like 430°C (800 °F). There had been strong support for a theory due to D. H. Menzel and F. L. Whipple, in America, that there were widespread seas, and that the clouds were made up chiefly of H_2O, but clearly liquid water could not exist under such conditions. Another surprise was that there seemed to be no measurable magnetic field; up to then it had been tacitly assumed that the field would be about the same strength as that of the Earth. Finally, the rotation period was very long indeed. Venus proved to be much less welcoming than Mars, or even the Moon, and the chances of finding any life there seemed to be ruled out. An astronaut standing on the surface would find himself in a gloomy, forbidding world, with the Sun hidden by the ever-present clouds and with an unbearable temperature. What living things could survive there? The answer seemed to be: "None".

Mariner 2 is still going round the Sun. Contact with it was eventually lost on 4 January 1963, at 14 hours GMT, 130 days after launch, by which time the distance from Earth amounted to as much as 54,000,000 miles. Certainly it had been a resounding success, even if much of the information had been unwelcome; one result of the mission was to relegate Venus to Target No.3 in the space programme, and promote Mars above it in order of priority. In the pre-1967 period the Americans launched no more Venus probes, though the Russians dispatched three.

Little need be said about Zond 1 of 1964, because it missed its target by a wide margin, and in any case it lost contact before it had produced much of value. (Not so with Mariner 2, which sent back data all through its outward flight; in particular, it produced useful information about the solar wind.) Then, in late 1965, the Soviets sent up two Venus probes within a week. Venera 2 by-passed Venus at 15,000 miles in the following February, but it too lost contact before the critical period, and was permanently lost.

Venera 3, one of the Soviet probes 1966.

It was Venera 3 which caused some excitement, and a great deal of controversy. Moscow reports claimed that on 1 March 1966 the probe actually landed on the surface of Venus.

The landing was not a controlled one, so that Venera 3 must have been destroyed on impact. Contact with the probe was lost before the final descent through the planet's atmosphere, so that no scientific data were received. On balance Venera 3, like its predecessor, must be classed as a failure. In the light of what we know now, it seems probable that the probe was literally crushed during the descent, because the atmosphere of Venus turned out to be much denser than had been expected.

In 1967 there were two launches, one Russian and one American. The Soviet Venera 4 was designed mainly to study Venus' atmosphere, though it was also planned to drop one small capsule on to the surface. Venera 4 entered the planet's atmosphere on 18 October, and achieved most of its aims; contact was maintained down to sixteen miles above the surface, and the

Venera 7, the first probe to transmit signals from the surface of Venus (1970).

capsule landed in the area we now call Eistla Regio, though nothing was heard from it after arrival. Mariner 5 flew past Venus in October at a range of 2,550 miles, and sent back useful data. (In case you are wondering about Mariners 3 and 4, both were Mars probes; Mariner 3 failed, but Mariner 4 was a real triumph, and sent back the first close-range views of the surface of the Red Planet.)

The main Soviet aim at this stage was to make a controlled landing on Venus, and receive pictures from the actual surface. What they did not appreciate was the unexpectedly high density of the Venus atmosphere, and Veneras 5 and 6, of May 1969, suffered the same fate as Venera 3 – the landers were crushed during the descent. Venera 7 (December 1970) was strengthened to cope with the tremendous pressure, and it worked. After entering the upper atmosphere, the landing capsule was jettisoned, and after aerodynamic braking a parachute was deployed. Venera 7 touched down in the area now called Navka

Mariner 10 was the first planetary mission to use the phenomenon of gravitational assist – using the gravity of one celestial body to change the spacecraft's speed and trajectory to aim it at a second celestial body. Mariner 10 was launched from Kennedy Space Center on November 13, 1973. Its closest approach to Venus – 3,000 miles – took place on February 5, 1974. Mariner 10 next encountered Mercury for the first time on March 9, 1974. Project scientists and engineers then realized that Mariner 10 could pass Mercury two more times. The second Mercury encounter occurred on September 21, 1974, and the third on March 16, 1975. Mariner 10 took more than 7,000 pictures of Venus, Mercury, Earth and the Moon.

Planitia, and sent back signals for 23 minutes after landing, so becoming the first man-made object to transmit from another planet. Venera 8 (July 1972) did even better, and after landing, again in Navka Planitia, transmitted for 50 minutes before losing contact. The high surface pressures and temperatures were fully confirmed, and it was also possible to measure the surface illumination, estimated to be about the same as that in Moscow on a cloudy winter day. It was also found that the windspeeds on the surface were very low.

The sole American effort around this time was Mariner 10, launched from Cape Canaveral on 3 November 1973. The main target was not Venus, but the innermost planet Mercury, which up to that stage had remained unvisited. En route, Mariner 10 passed Venus at a range of 3,600 miles, sending back good images of the cloud-tops as well as obtaining useful data; it used Venus' gravitational pull to send it on its way to Mercury, which was encountered in March 1974. Two more passes of Mercury were made, on 21 September 1974 and 10 March 1975, and excellent pictures were received of the craters on that bleak little planet. Indeed, almost all we know about the surface of Mercury is due to Mariner 10.

So far as Venus was concerned, the next major advances came in 1975, with two more Soviet spacecraft: Veneras 9 and 10. Both were dispatched in the summer, and landed on 22 and 27 October respectively, about 150 miles apart. Before starting their descent through Venus' atmosphere they were chilled below freezing-point; after arrival each operated for about an hour (to be precise, 53 minutes for Venera 9 and 65 minutes for Venera 10), and each sent back a picture of the surface, relayed to Earth by the orbiting sections of the probes which had been left in closed paths around Venus.

The pictures were of amazingly good quality. That from Venera 9 showed a rock-strewn landscape; the rocks were sharp-edged and most of them were from 2 to 4 feet across. One Soviet astronomer, M. Marov, described the scene as "a stony desert", and there was no marked evidence of erosion, which was surprising in view of the fact that even a slow wind would have tremendous force in the thick atmosphere of Venus. Measurements indicated a windspeed of about 7 knots, but this would be as devastating as the effect of a sea-wave on a terrestrial cliff during a 40-knot gale. The scene from Venera 10 was much the same, but it seemed that the large rocks were outcrops from below the surface rather than simple scattering, as on the Venera 9 site. Also, the rocks were rather smoother, and the general outlook was of a landscape which was appreciably older.

The first views of Venus' surface, sent back from Veneras 9 and 10, in October 1975.

There had been earlier suggestions that the surface of Venus might be too dark to allow any picture-transmission at all. Both the Veneras took floodlights with them, but there was no need to use them. It had also been thought that the great density of the atmosphere might bend light-waves strongly enough to produce what is called "super-refraction", so that an observer on the surface would seem to be standing in a huge bowl, with the horizon curving upward on all sides of him. This would have been weird in the extreme, but it proved not to be the case.

Veneras 11 and 12 (December 1978) were of essentially the same type as their predecessors, carrying both orbiters and landers. The instruments had been improved, and both probes sent back signals after landing, but the television cameras failed to work. Meanwhile, the Americans had produced their most ambitious probes yet – Pioneers 12 and 13, otherwise known as Pioneer Venus 1 and Pioneer Venus 2.

Pioneer 12 was a pure orbiter. It was launched on 20 May 1978, and was programmed to enter a closed orbit round Venus on the following 4 December, which was also the scheduled arrival time

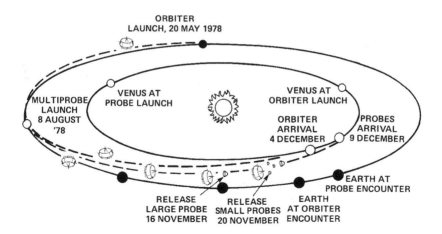

The Pioneer Venus mission - flight paths of the Venus 'armada' of 1978.

for Pioneer 13. Its path round Venus was very elliptical, taking it from its closest point to the planet, a mere 93 miles above the surface, out to 91,000 miles; the orbital period was just over 23 hours. Observations were made of the upper atmosphere, clouds were recorded and temperatures measured; the Venus surface was mapped by radar, and, *en passant*, systematic observations were made at ultra-violet wavelengths of several comets, including Halley's. Pioneer 12 operated for a very long time, and did not burn up in Venus' atmosphere until 9 October 1992.

Pioneer 13 was even more ambitious. It was a multi-probe mission, consisting of a "bus" carrying four landers; the so-called Large Probe, 5 feet in diameter, and three identical smaller vehicles, the North Probe, the Day Probe, and the Night Probe. The whole assembly was launched from Cape Canaveral on 8 August 1978. Three months later, on 16 November, the Large Probe was released from the Bus, and began its independent journey to Venus; it was aimed at the volcanic area of Beta Regio, in the northern hemisphere. Four days later the other three probes were released; two were aimed at the daylit side of Venus, at Ishtar Regio and Themis Regio respectively, and the third was to come down on the night hemisphere, north of Aino Planitia. On separating from the Bus they were powered down, because

The first color photograph of Venus' surface, returned from the Venera 13 probe in 1982.

they had no solar cells to recharge their batteries, but were programmed to become active again three hours before landing. All four sent back valuable data during their last moments. They were not designed to survive the landing on the surface, though in fact the Day Probe did so, and transmitted for 67 minutes after arrival. The Bus had no braking mechanism, and simply crash-landed in Themis Regio.

Both Pioneers had been extremely successful. It was confirmed that the surface temperature is high enough to melt lead, and that the atmosphere is almost pure carbon dioxide (97 per cent) together with a little nitrogen and trace amounts of other gases. The clouds are rich in sulphuric acid, and though they were not so thick as had been often supposed, they were quite capable of hiding the surface permanently.

Veneras 13 and 14 (1981) were identical; each consisted of an orbiter and a lander, and both were successful, carrying out the first attempts at soil analysis. Each transmitted for a full hour after arrival, but it had become clear that no spacecraft, however durable, could remain active for long under those hellish conditions.

The next two probes, both Russian, were of different type.

Their main mission was to study not Venus, but Halley's Comet!

Comets, as we have noted, are the most erratic members of the Solar System. Most of them (not all) have very eccentric orbits, and, with one exception, all the really bright comets have revolution periods of many centuries. The exception is Halley's Comet, which returns every 76 years. It was on view in 1835 and 1910, and was due back once more in 1986. Astronomers were very much on the alert, and plans were made to send space-craft to make contact with the comet. Eventually, five probes were launched: two Russian, two Japanese and one European. The Americans pulled out on the grounds of expense – which they soon regretted, and will continue to regret until the comet comes back once more in 2061.

The Soviet vehicles were known as Vega-1 and Vega-2. (This has absolutely nothing to do with the brilliant blue star Vega, or Alpha Lyrae, which as seen from Britain is almost overhead during summer evenings. The name comes from *Ve*nus and *Ga*llei; Gallei is the Russian name for Halley, since there is no letter H in the Russian alphabet.) The probes were launched on 15 and 20 December 1984 respectively, and in June 1985 were within range of Venus. Each released a lander; that of Vega 1 entered Venus' atmosphere on 11 June, and the second lander followed four days later. Each came down in the region of Rusalka Planitia, not far from the Venus equator, during darkness; each transmitted for 20 minutes after touch-down. This was not all. During the descent, each probe released a balloon. The balloons drifted around at various levels, at an average of about 30 miles above the ground, and sent back data about windspeeds and temperatures; on Earth, twelve stations at widely separated sites took part in tracking. Finally the balloons drifted on to the day side of the planet, where they expanded because of the rise in temperature and then burst. The Vega probes themselves used the gravitational pull of Venus to send them on to a rendezvous with Halley's Comet in March 1986. The two Japanese spacecraft and the European vehicle, Giotto, made successful rendezvous with Halley's Comet, but went nowhere near Venus en route.

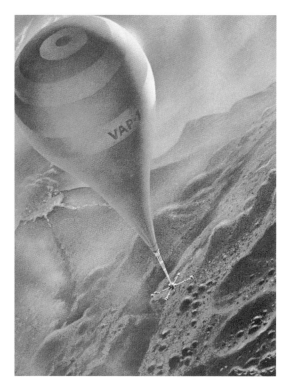

Impression of the balloon dropped into Venus' atmosphere
from the Soviet probe Vega 1 (1985).

We come now to Magellan, without doubt the most successful Venus probe to date. It was launched in May 1989, entered Venus' orbit in August 1990, and continued making radar surveys of the surface until burning away in the Venus atmosphere on 11 October 1994. In little over four years Magellan completely revolutionized all our ideas about Venus as a world.

Magellan was a radar mapper. It was carried aloft in the Space Shuttle *Atlantis,* from Cape Canaveral; when *Atlantis* was in low Earth orbit, Magellan was released from the cargo bay. A solid-fuel motor then fired, and sent the probe on its way, orbiting the Sun one and a half times before nearing Venus. A second solid-fuel motor then fired, putting Magellan into a closed path round the planet. The initial orbit was highly eccentric, and the distance from Venus ranged between 182 miles out to 5,296 miles in a period of 3.2 hours. The orbit took Magellan over the poles; later the orbit was modified several times.

Magellan was 15.4 feet long. Topped by a 12-foot high-gain antenna, launch weight was 7,612 pounds. It was powered by two square solar panels, each with a side of $8^1/_2$ feet. Obviously these became less effective as the mission progressed, but they lasted for as long as was necessary. In the end about 98 per cent of the total surface of Venus was mapped by radar.

Magellan was an unqualified success, and scientists were sad when it came to the end of its career. Since then two more spacecraft have by-passed Venus; Galileo in 1990, on its way to Jupiter, and Cassini, which was bound for Saturn and encountered Venus twice, once in April 1998 and again in June 1999, before starting the final leg of its long journey to the Ringed Planet. Both provided useful information, but it is fair to say that almost all we know about the nature of Venus is due to the Russian landers and, above all, to Magellan.

Patrick Moore with a model of Magellan.

11

The Landscapes of Venus

Because telescopes can never penetrate the Venus clouds, all maps have to be obtained by the use of radar. Earth-based radar equipment, mainly at Goldstone in the United States and Arecibo in Puerto Rico, gave preliminary results, but we depend chiefly on Magellan. The spacecraft's main "dish", just over 12 feet across, sent pulses downward at an oblique angle to the probe itself, striking the surface below just as a beam of sunlight will do on the Earth. The surface rocks modified the pulse before reflecting it back to the antenna; in radar, rough areas look bright while smooth areas are dark. A smaller antenna sent down vertical pulses; the time-lapse between transmission and "echo" gave the height of the surface to an accuracy of less than 100 feet – and contour maps were obtained of 98 per cent of the entire planet.

In one way the usual maps give a false impression, because low-lying areas are usually coloured blue, while higher regions are shown in yellow and red. Of course these colours are false, and are put in only to make the maps easier to interpret, but it is only too easy to imagine that the blue areas are seas, or at least old sea-beds. This is quite wrong. There are no seas on Venus now; there may once have been, but there is absolutely no connection between these hypothetical seas and the blue areas on the maps. In fact, about 60 per cent of the surface lies within 1,600 feet of the mean radius of Venus, with only 5 per cent more than a mile and a quarter above it; the total elevation range is 8 miles, from the Maxwell Mountains, 6.8 miles above the mean radius, and the bottom of the Diana Chasma trench, 1.2 miles below it. Venus,

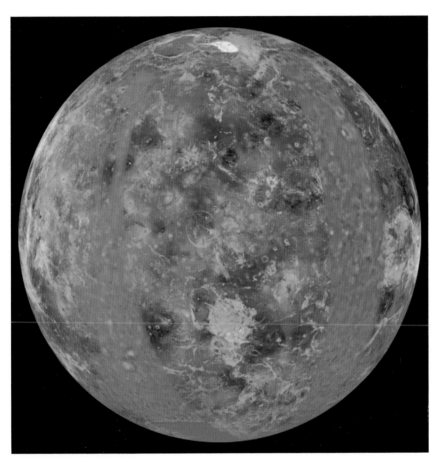

Venus without clouds.

incidentally, is almost a perfect sphere. On Earth, the equatorial
diameter is 27 miles more than the diameter as measured through
the poles, but on Venus the polar and equatorial diameters are
virtually the same.

For convenience, the various features on Venus have to be

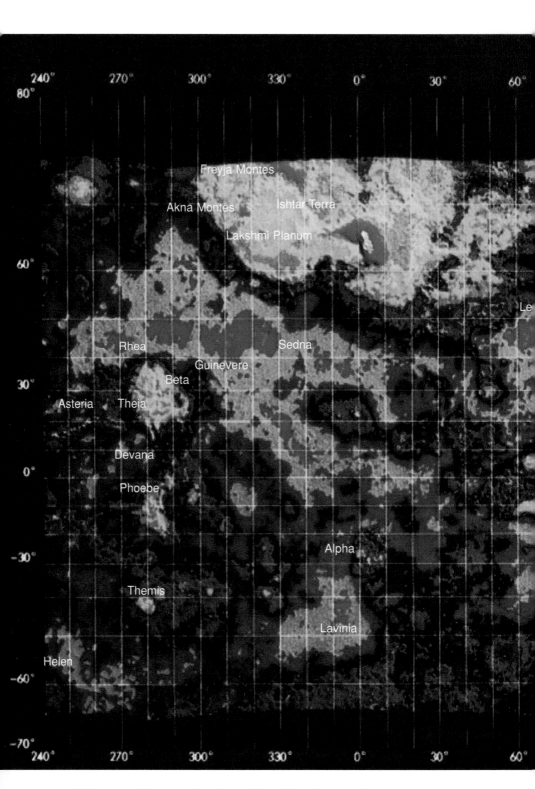

Radar map of Venus – a Mercator projection of the planet's surface.

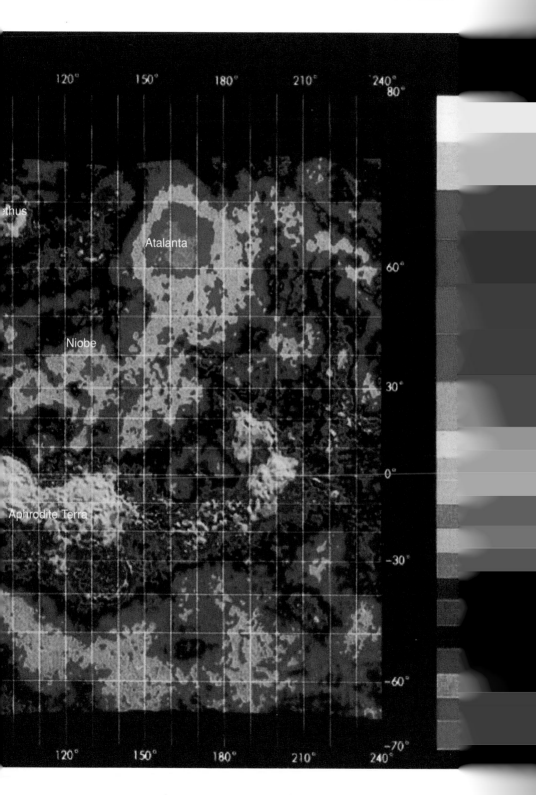

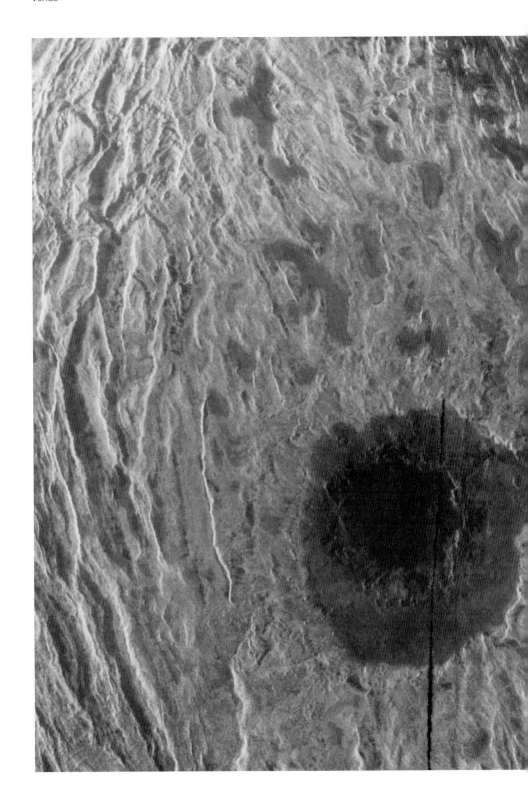

named, and this is the responsibility of the Nomenclature Committee of the International Astronomical Union, of which I have the honour to be a member. To be Politically Correct, it was decreed that all names on Venus must be female. I may add that this edict was laid down before I joined the Committee, and moreover the name of James Clerk Maxwell, the great Scottish physicist, had already been allotted to the highest mountain peaks on Venus; it had been identified by Earth-based radar. (Maxwell is still there, and there is nothing that the Politically Correct lobby can do about him.) We have to admit that some of the names are unfamiliar. Everyone will know Boadicea, Florence Nightingale, Marie Curie and Kirsten Flagstad, but what about Vires-Akka, Vellamo and Xochiquetzal? In fact Vires-Akka is a Lapp fertility goddess, Vellamo a Finnish mermaid, and Xochiquetzal the Aztec goddess of flowers. Moderns are not neglected; for example, craters are named after the Russian painter Bashkirtse, the American social worker Jane Addams (no relation to the television family) and the Egyptian authoress Al-Taymuriyya. If permanent bases are ever set up on Venus, which admittedly seems rather dubious, there will be some interesting postal addresses!

The surface of Venus is divided into highlands, volcanic plains (planitiae) and the vast rolling plains. The highlands account for about 8 per cent of the surface, the planitae for 27 per cent and the upland rolling plains for the remaining 65 per cent. Vulcanism dominates the entire surface, and there are lava-flows everywhere; nothing could be further removed from the pre-Space Age picture, when it was often thought that a probe could splash down quite comfortably in a warm, welcoming ocean.

The highland "regiones" (regions) seem to have been formed by upswelling from below the planet's crust; large shield volcanoes, of the same basic type as those in Hawaii, are found on them. There are two major highlands, often referred to, misleadingly, as continents: Ishtar Regio in the northern hemisphere, and

Cleopatra, a 12-mile crater on the eastern slopes of Maxwell Montes. Its origin is uncertain (NASA).

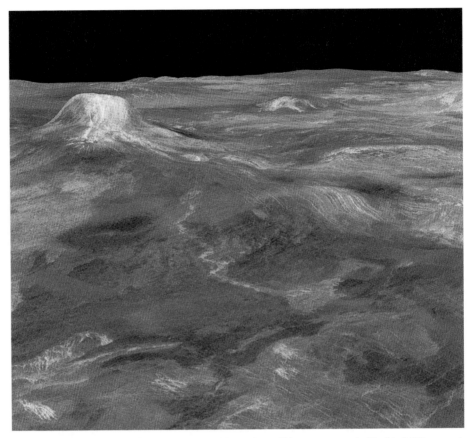

Lava plains and volcanoes dominate the surface of Venus, as revealed in this 3-D image of Eistla Regio and Gula Mons generated from Magellan radar data (NASA).

Aphrodite Regio near the equator. There are two smaller highland "terrae", Lada Terra in the south, and a less well-defined feature encompassing Beta Regio, Phoebe Region and Themis Regio.

Ishtar Terra is some 3,100 miles across, which is about the size of Australia. It consists of western and eastern components separated by the Maxwell Mountains (Maxwell Montes), the highest on Venus. At the top, more than 39,000 feet above the mean radius, is a 66-mile crater, Colette, once thought to be volcanic though its

Mona Lisa (right), a 53-mile-wide crater. The mountains around the rim were raised during the impact, and the crater is surrounded by a blanket of ejecta (NASA).

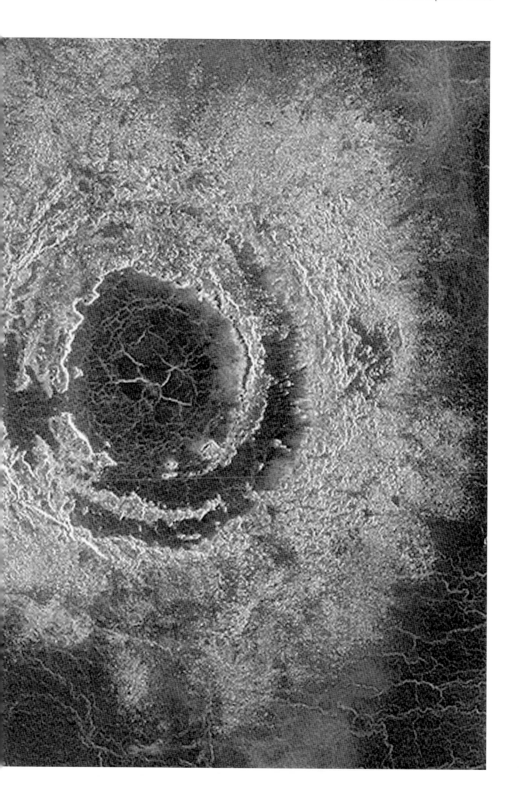

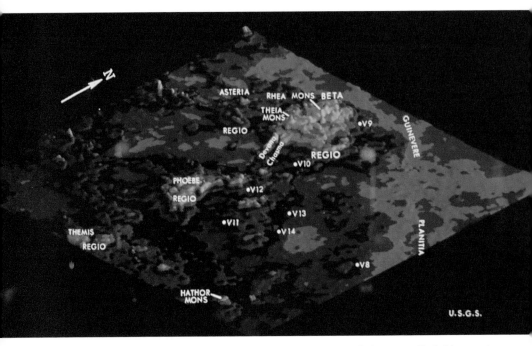

Radar map of Beta Regio showing the positions of its two main features – Theia Mons and Rhea Mons.

precise nature is rather uncertain. The steep-sloped Maxwell Mountains certainly dwarf our own Himalayas, even if they cannot match the colossal volcanoes of Mars. The adjoining Lakshmi Planum is from 13,000 to 16,000 feet above the main radius, about the same level as the Tibetan plateau on Earth; it is lava-covered, and is bounded by the Danu, Freyja and Maxwell Mountains. To the south-west it is bounded by the steep Vesta Rupes escarpment, with other escarpments to the south-east and north-east.

The other important highland, Aphrodite Terra, lies mainly in the southern hemisphere, though it does just straddle the equator. It is not so lofty as Ishtar, and rises to only about 16,500 feet, but it is distinctly larger; its area is roughly the same as that of Africa. Western Aphrodite contains two high plateaux, Ovda Regio and Thetis Regio, both between 10,000 and 13,000 feet in altitude. Ovda is a region of "tessera" structure, cut by ridges and troughs; this sort of terrain was once known as "parquet", but the term was rejected as being insufficiently scientific! The slopes of Thetis Regio are slightly less steep than those of Ovda, and the plateau is somewhat concave.

The central part of Aphrodite is cut by ridges and troughs. Here we find the very deep, impressive Diana Chasma and Dali Chasma; the bottom of Diana Chasma is actually the lowest point on the surface of Venus, 6,500 feet below the mean radius. Eastern Aphrodite is known as Atla Regio, another plateau characterised by massive volcanoes: Ozza, Maat, Sapas and others. Maat Mons, more than 16,000 feet high with a 125-mile diameter base, is crowned by several calderae, with extensive lava-flows. The shape of this part of the continent has led to its being nicknamed "the Scorpion's Tail".

About 1,900 miles north of Ovda Regio lies a high plateau, Tellus Regio, with a surface structure apparently much the same as with Ovda and Thetis.

The third important highland area is Beta Regio, which lies mainly – not entirely – south of the equator, adjoining Phoebe Regio and Themis Regio. Beta is of special interest because it includes two huge structures, Theia Mons and Rhea Mons. Both these were once thought to be volcanoes; this is true of Theia, which is indeed the most massive volcano known on Venus, but not of Rhea, which is, however, scored by volcanic deposits. Theia is over 15,000 feet high, with a base diameter of 220 miles and an oval central caldera measuring 46 miles by 30 miles. Lava-flows from Theia extend over an area more than 500 miles wide; the volcano lies at the junction of three rifts, one of which, Devana Chasma, is 125 miles wide in places, making our Arizona Canyon seem very puny indeed. South of Rhea, between Beta Regio and Phoebe Regio, the chasm is almost 4 miles deep. Its total length is just over 1,000 miles.

There is one more notable highland, Lada Terra in the south; it is bordered by Helen, Lavinia and Aino Planitiae. Alpha Regio lies 1,350 miles north of Lada, and is joined to it by a system of deep rifts with raised rims.

An extensive lowland area shaped like a letter X lies south of Ishtar Terra, between Aphrodite Terra and Beta Regio. The region,

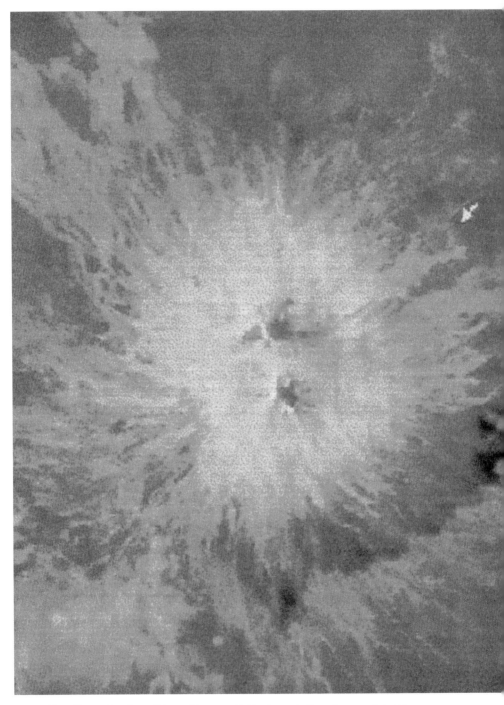

Magellan radar view of Sapas Mons, a shield volcano built up from overlapping lava flows (NASA).

known as Guinevere Planitia, in the north-west of the X, is the most rectilinear of the lowland areas. The other lowland regions of the planet are quite isolated, and form discrete patches among the other regions of the Venus surface.

East of Ishtar Terra the surface descends to the most extensive basis on the planet, Atalanta Planitia, centred at latitude 65°N, longitude 165°. Atalanta is about the size of the Gulf of Mexico, and lies about three-quarters of a mile below the mean radius; the lowest portion goes down to a full mile, nearly as low as Diana Chasma. Atalanta is comparatively smooth, resembling the lunar Mare basins, the Martian northern plains, and the ocean basins of Earth. No large craters are found in these lowland regions, so that they may be basaltic lava-flows similar to those which fill the Mare basins on the Moon. Certainly this idea is supported by the fact that the rocks near Beta Regio are essentially basaltic. Moreover, the crust is thinner and less dense under the plains than below the upland regions.

Thousands of volcanoes have been identified on Venus, ranging from tiny vents to huge shields hundreds of miles across. They are least common in the tessera ("parquet") regions, where there are linear ridges and troughs. Shield fields, each with many small volcanoes, are widespread. The large shield volcanoes are of the same type as Mauna Loa in Hawaii, but they are much more massive, and are also much higher. There is a good reason for this, as we now know.

On Earth, the crust drifts around over the mantle – as was first realized by the Australian meteorologist Alfred Wegener, who had to wait for many years before his theory of "plate tectonics" was accepted. Hawaii provides an example of the way in which the theory works. A volcano is formed over a "hot spot" in the underlying mantle, and erupts for a definite period before it drifts away from the hot spot and ceases to be active. Mauna Kea, on which some of the world's most powerful telescopes have been set up, was active thousands of years ago, but has now moved away from the spot, to be replaced by its neighbour, Mauna Loa, which is very active indeed. In several thousands of years Mauna Loa also

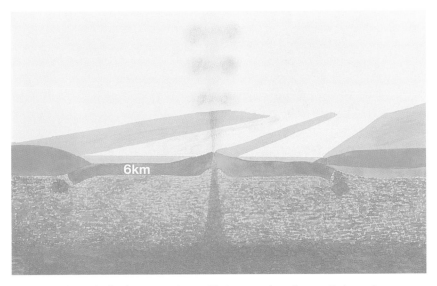

6km

Structure of a typical volcano – a plume of hot magma from the mantle forces its way up through the planet's crust and eventually erupts, releasing gas and lava which builds up into the classic cone shape.

will drift away, and become dormant. But with no plate tectonics on Venus, a volcano remains over its hot spot for a much longer time, and can build up to immense size.

Theia Mons is a case in point. Another splendid example of a shield volcano is Gula Mons, in the western part of Eistla Regio. It is almost two miles high, and is a member of a chain of volcanoes, which also includes Sif Mons, 430 miles from Gula and almost as lofty. Western Eistla Region is a broad dome, almost a mile above the mean radius of Venus, and 1,250 miles wide.

A third splendid example of a shield volcano is Maat Mons, almost on the equator in the Scorpion's Tail area. In size and elevation it is almost the equal of Theia, and if it could be seen from the surface of the planet it would indeed be impressive. Sapas Mons, rising from Atla Regio, is a mile high, and is characterised by the two smooth, flat-topped areas at its double summit (remember that in radar, these smooth areas appear dark). Lava-flows round Sapas are extensive, and many of the flows seem to have been erupted along the sides of the volcano rather than from

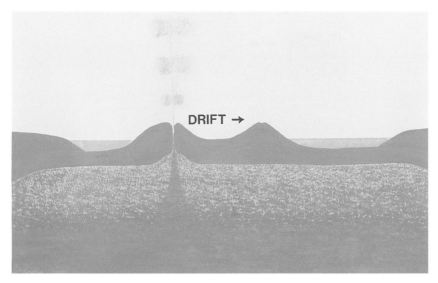

A volcanic island chain such as the Hawaiian Islands, forms as Earth's continental drift carries a surface plate across a hot spot in the mantle.

Structure of Beta Regio – Venus' crust seems to be far thicker than Earth's, and the lack of tectonics means that volcanic lavas can pile up to form truly huge structures.

the summit – as is the case with many terrestrial volcanoes.

The volcanoes of Venus are of many different types. There are "anemone" shields, where the volcano is ringed by narrow, radar-

Sapas Mons and Maat Mons. Seen in the foreground of this computer-generated, three dimensional perspective view of the surface of Venus is the volcano Sapas Mons – 248 miles wide – on the western edge of Alta Regio. Sapas Mons, named for a Phoenician goddess, is 0.9 mile high, with a peak that sits at an elevation of 2.8 miles above the planet's mean elevation. In the background on the horizon is Maat Mons, the largest shield volcano on Venus. Its peak rises to an elevation of 5 miles above the planet's mean elevation. Lava flows extend to the base of Sapas for hundreds of miles across the plains shown in the foreground (NASA).

bright lava-flows. There are "ticks", which have broad bases and somewhat concave summits. There are what are called "pancake domes", which are not particularly high; they are always circular, and may be due to very thick lava-flows breaking through the surface from below. A particularly good chain of pancake domes is found near the eastern edge of Alpha Regio; there are several components, averaging 15 miles in diameter and 2,500 feet in height. Some volcanoes are very irregular in form; one of these, in Aino Planitia, has a complicated structure, suggesting that the erupted lava was unusually viscous.

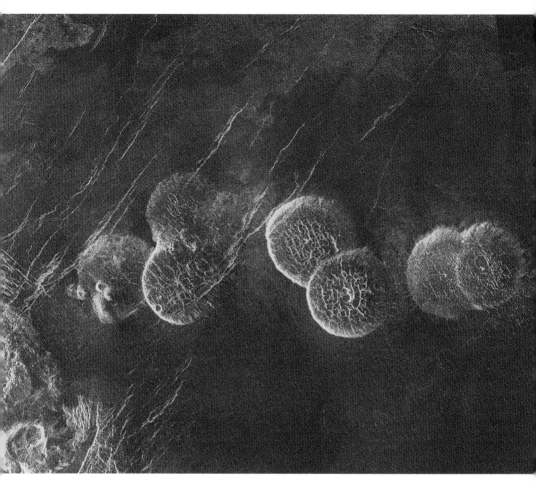

"Pancake Domes" on the eastern edge of Alpha Regio. Seven circular, dome-like hills, averaging 15 miles in diameter with maximum heights of 2,475 ft, dominate the scene. These features are interpreted as very thick lava flows that came from an opening on the relatively level ground, which allowed the lava to flow in an even pattern outward from the opening. The complex fractures on top of the domes suggest that if the domes were created by lava flows, a cooled outer layer formed and then further lava flowing in the interior stretched the surface. The domes may be similar to volcanic domes on Earth. Another interpretation is that the domes are the result of molten collapse and fracturing of the dome surface. The bright margins possibly indicate the presence of rock débris on the slopes of the domes. Some of the fractures on the plains cut through the domes, while others appear to be covered by the domes. This indicates that active processes predate and postdate the dome-like hills. The prominent black area in the northeast corner of the image is a data gap. North is at the top of the image (NASA).

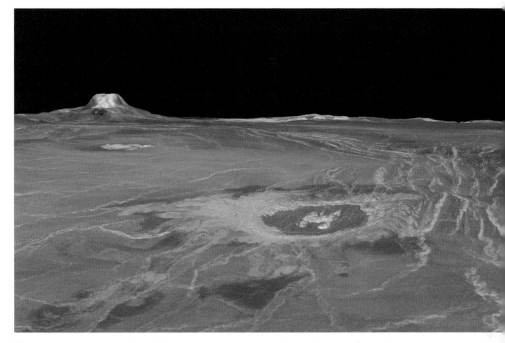

The 2-mile high volcano Gula Mons rears up behind the 30-mile diameter Cunitz impact crater. In this and all the Magellan 3-D images, the vertical scale has been exaggerated many times (NASA).

Are the Venus volcanoes active now? Most astronomers believe the Martian volcanoes to be dead, but the situation on Venus is very different. The surface is young by cosmical standards – probably no more than 500 million years old – and seems to be in a constant state of re-surfacing, which indicates current vulcanism. Also, it has been found that the amount of sulphur dioxide in the atmosphere changes, and shows occasional sharp increases as though gas had been released from an erupting volcano. The jury is still out, but all in all it does seem very probable that violent volcanic activity is going on today.

Other tectonic features include coronae, arachnoids and novae. Coronae are unique to Venus, at least as far as we know. Hundreds of them have been identified; most are around 150 miles across, though some are considerably larger. A typical corona is a circular structure, with a circumference defined by a raised ring-like zone

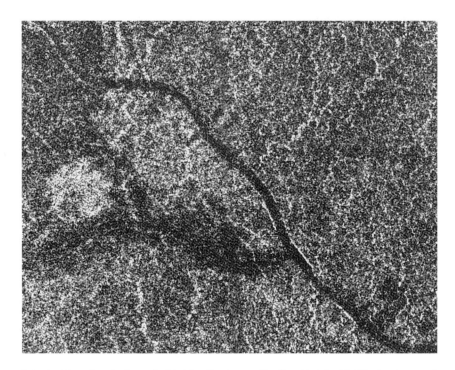

A typical lava flow is shown in this Magellan image. The flow (dark line) is 20 miles long and up to 2 miles wide in places. It probably formed from relatively fluid lava – the opposite of the viscous lava which probably formed the pancake domes (NASA).

made up of ridges and troughs. It is thought that a corona may be due to a mass of hot magma from below the surface rising upward and breaking the crust, causing it to melt and collapse, generating volcanic flows and fault patterns radiating from the centre of the structure, which may be either slightly above or slightly below the main radius of Venus. Arachnoids, which may be up to 140 miles across, are so named because of their spidery and cobweb-like appearance in the radar images. A typical arachnoid has a central volcanic feature surrounded by a network of fractures; in some ways they are not unlike coronae, though they are much smaller.

Dickinson Crater (right) an impact crater 43 miles across. The centre has flooded with dark material and it is surrounded by ejecta except to the west – an indication that the impactor may have struck from that direction. Such "lopsided" craters are unique to Venus (NASA).

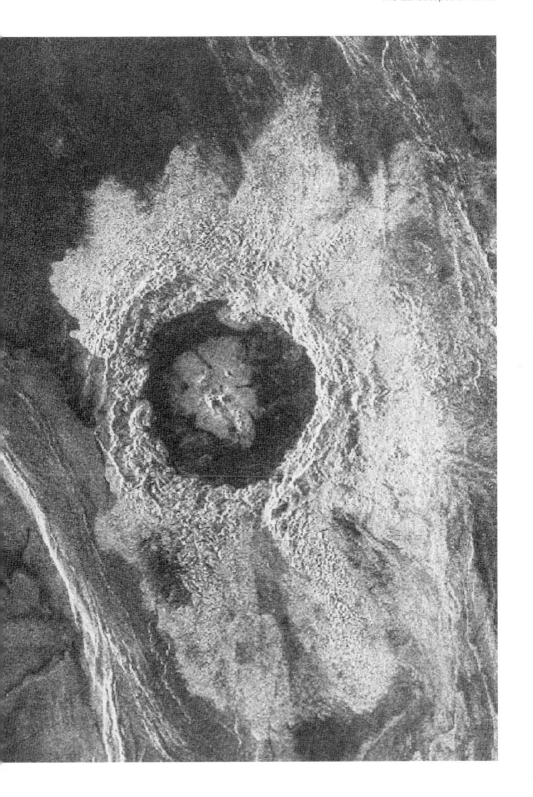

A highly fractured dome, 78 miles long, on the eastern flanks of Freyja Montes. The dome appears to be a local region that has been uplifted and pulled apart within a region of mountain belts. The "turtleback" appearance of the dome is the result of two sets of intersecting fractures, one that extends roughly north to north-northeast and the other west to west-northwest. Individual cliffs or slopes appear to define troughs, called graben, commonly 0.6 by 3 miles wide. In the eastern portion of the image, the fractured terrain has been covered by volcanic deposits. The appearance of a feature that has been pulled apart, in a region that has been compressed is important in studying how mountains are built on Venus. North is at the top of the image (NASA).

The Lavinia Planitia craters (left) range in size from 23 to 31 miles across, and are typical of many on Venus. The image covers an area 342 miles wide by about 311 miles long. Situated in a region of fractured plains, the craters show many features typical of meteorite impact craters, including rough (bright) material around the rim, terraced inner walls and central peaks. Numerous domes, probably caused by volcanic activity, are seen in the southeastern corner of the mosaic. Some of the domes have central pits that are typical of some types of volcanoes. North is at the top of the image (NASA).

Novae also show a radial structure, and it has been suggested that they represent early stages in the formation of coronae.

It is obvious that vulcanism has been (and probably still is) the dominant process on Venus, and there are volcanic craters, often with central peaks. But there are also impact craters, such as Dickinson in the north-east part of Atalanta; here we have a 43-mile crater, with an incomplete central ring and a floor which is partly radar-dark (smooth) and partly radar-bright (rough). Impact craters are much less common than they are on Mercury or the Moon. There are two main reasons for this. First, small meteoroids are burned away in Venus' dense atmosphere before they can reach the ground. Secondly, the atmosphere slows down many of the larger meteoroids to such an extent that they land with insufficient force to produce craters. There is also the constant re-surfacing, so that old craters have been eroded away, and all traces of the "Great Bombardment", which scarred the faces of the Moon and Mercury, have been obliterated.

We do not yet know a great deal about the composition of the rocks, but in view of the dominant vulcanism it is reasonable to suppose that they are generally basaltic, as with volcanic rocks on Earth. For full information we must await the time of a sample-and-return probe, capable of collecting material and bringing it home for analysis. This should soon be possible for Mars, but obtaining material from Venus is going to be so difficult that we will probably have to wait for a good many years.

Venus is certainly a strange, unfriendly place. Yet it is a fascinating world, and no doubt it has many more surprises in store for us during the present century.

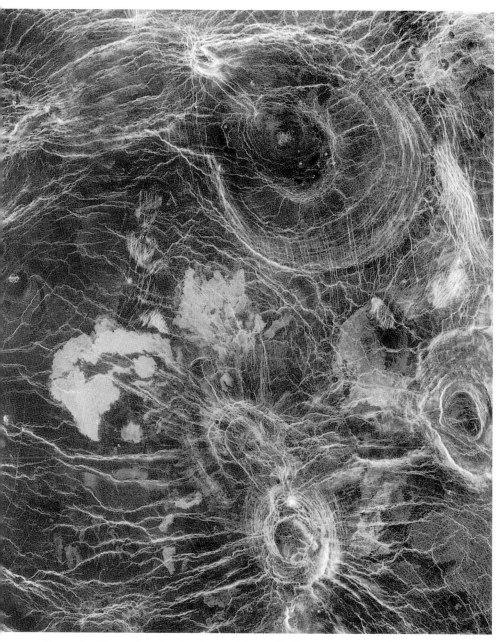

The spider-like arachnoids in this image of Sedna Planitium range between 31 and
148 miles in diameter. They seem to be the result of subsurface magma forcing the crust
upward and creating complex fault systems. The bright patches in the centre of the image
are lava forms (NASA).

12

Structure and Atmosphere

In size and mass, Venus and Earth are so alike that we may well assume that their internal structures are much the same. This may not be so very wide of the mark, but there are important differences. Venus is appreciably less massive than our world, and it is also less dense. Moreover it has no detectable magnetic field, which affects our ideas about the composition of the inner core; and of course it is a slow rotater.

We know a great deal about the make-up of the Earth, mainly from studies of earthquake waves. These are of three main types: primary ("push" waves), secondary ("shake" waves), and surface waves. Primary waves can travel through liquids, but secondary waves cannot, and it was this which gave the first definite proof that the Earth's core must be liquid – though the innermost part is now known to be solid. The rigid outer crust and the upper mantle of the Earth's globe make up what is termed the lithosphere. The crust has an average depth of 6 miles below the oceans, but up to 30 miles below the continents; underneath comes the mantle, which extends down to 1,740 miles and accounts for 67 per cent of the total mass of the globe. The outer liquid core extends down to 3,200 miles, and is hot; the temperature is given as around 4,500 Celsius(8,100°F). Finally there is the solid innermost core. Movements in the iron-rich liquid layer are responsible for the Earth's magnetic field.

The structure of Venus may be similar, but the lack of an overall magnetic field indicates that the liquid core is smaller than ours, not only absolutely but also relatively. The Venus lithosphere is

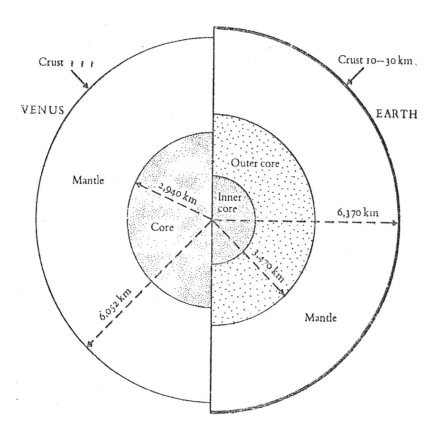

Crust ？ ？ ？

VENUS

Mantle

Core

2,940 km

6,052 km

Crust 10–30 km

EARTH

Outer core

Inner core

6,370 km

3,470 km

Mantle

Diagram showing the interiors of Earth and Venus compared.

certainly basaltic, and may go down to at least 40 miles in some areas, notably below the tesserae – but we cannot be at all definite, because we cannot make use of "quake" waves even if they occur

Venus volcano, painting by Paul Doherty.

(as they probably do). When we manage to set up an automatic recording station on Venus and measure the ground tremors we will undoubtedly learn a great deal more, but this lies some way in the future. So far as the core temperature is concerned, all we can really do is to make an educated guess; it may well be that the temperature at the centre of the globe is rather less than that of the Earth.

With the atmosphere we are better placed, because we can study it directly. The first really useful investigations were made from 1923 to 1928 by E. Pettit and S. B. Nicholson, using the 100-inch reflector at Mount Wilson in California. They found that the temperature of the upper clouds was -38°C (-36°F) for the day side and -33°C (-28°F) for the night side, which is in excellent agreement with modern results, though at that time the axial rotation period was not known (note, incidentally, that because the rotational axis is almost perpendicular to the orbital plane, there are no true "seasons" on Venus). Instead of the expected oxygen and nitrogen, Pettit and Nicholson found that the main constituent of the upper atmosphere was carbon dioxide. This is a heavy gas, and would be expected to sink rather than rise – so that it had to be assumed that the atmosphere was mainly carbon dioxide all the way down to the surface, and the greenhouse effect would lead to a very high surface temperature. In fact the atmosphere is made up of 96 per cent carbon dioxide, 3.5 per cent nitrogen, and mere traces of other gases such as argon, oxygen, sulphur dioxide and carbon monoxide, though – as we have noted – the amount of sulphur dioxide shows marked and sometimes quite abrupt variations.

There was considerable discussion about the nature of the clouds. They could hardly be similar to our own clouds, because the amount of water in Venus' atmosphere is vanishingly small (no more than 0.0001 per cent at most). For some time it was thought that they might be due to a chemical substance called formaldehyde, but the water-droplets idea still had its supporters until shortly before the results were received from Mariner 4. We now know that the clouds are rich in sulphuric acid, and at some levels there must be a sulphuric acid "rain", though all of it will evaporate before reaching the surface of the planet. There are various cloud layers. The upper clouds lie at around 43 miles above ground level; the cloud deck ends at 18 miles above the surface, and below this the visibility is unexpectedly good, which is why the first Russian landers had no need to use the searchlights with which they had been equipped.

Though many of the problems set to us by Venus have been solved, others remain. Why is the direction of rotation opposite in sense to those of the Earth and Mars? Why do the upper clouds rotate so quickly, while the solid body of the planet is so leisurely?

The atmospheric circulation of Venus is revealed by this high-contrast image from Pioneer Venus. The planet's slow rotation means that the coriolis effect that complicates Earth's weather has no effect here, and the entire atmosphere functions as a single "Hadley Cell", transferring hot gases from the equator to the poles by convection. Onto this is overlaid a high-speed wind that carries Venus' clouds around the planet every four Earth days. (Mariner 5)

Why is there no magnetic field – and why is the surface so unbearably hot, hotter even than that of Mercury? The answers must lie in the fact that Venus is more than twenty million miles closer to the Sun than we are.

It is safe to assume that Venus and the Earth are of the same age, 4.6 thousand million years, and that they accreted from the solar nebula in the same way. But in those far-off times the Sun was not nearly so luminous as it is now, and it may well be that Venus and the Earth started to evolve along similar lines; oceans may have formed on Venus just as they did here. But then, quite quickly on the cosmical timescale, the Sun became more powerful. Earth, at a range of 93,000,000 miles, was out of harm's way; Venus, at only 67,000,000 miles, was not. The effects were devastating. The Venus oceans boiled away, the carbonates were driven out of the rocks, changing the atmosphere completely, and there was a runaway greenhouse effect, transforming Venus from a potentially life-bearing world into the inferno of today.

It is sobering to reflect that if the Earth had been only a few million miles closer to the Sun than it actually is, it would have suffered the same fate. Any higher forms of life which had managed to develop would have been wiped out – and you would not now be reading this book.

13

Life on Venus?

Venus is a scorching-hot world, with a crushing atmospheric pressure and acid clouds. Advanced life of our kind is out of the question there, and, as I have commented, any astronaut who lands and then steps out of his spacecraft will be promptly fried, poisoned, squashed and corroded. But this was not always the view among astronomers – such as Bernard de Fontenelle, a French Jesuit who was born in 1657 and lived to the advanced age of one hundred. By the standards of the time he was an excellent scientist, and from 1697 became Permanent Secretary of the Académie des Sciences in Paris. His most famous book, the English title of which is *The Plurality of Worlds*, was published in 1686; it takes the form of a conversation between a layman and an extremely well-informed if imaginary Countess. Fontenelle believed that all worlds were inhabited, and this is what he (or rather, his Countess) has to say about Venus:

> '*I find it is enough to guess at the inhabitants of Venus; they resemble what I have read of the Moors of Grenada, who were little black People, scorched with the Sun, Witty, full of Fire, very Amorous, much inclined to Musick and Poetry, and ever inventing Masques and Tournaments in Honour of their Mistresses.*'

Fontenelle is also able to provide information about the inhabitants of Mercury:

Venus in the crescent stage 1959 28-inch O. G. (Herstmonceux) x 600.

Drawing by Patrick Moore.

'They are yet nearer to the Sun, and are so full of Fire, that they are absolutely Mad; I fancy they have no Memory at all, no more than most of the Negroes; that they make no Reflections, and what they do is by sudden Starts, and perfect Hap-hazard; in short, Mercury is the Bedlam of the Universe.'

Another supporter of life on Venus was eighteenth-century Swedish philosopher Emanuel Swedenborg. He carried out a great deal of valuable scientific research, but with regard to extraterrestrial life he drew his information from personal conversations with angels and spirits, so that his conclusions seem hardly relevant here.

Coming on to more modern times, it is worth quoting the words of Camille Flammarion, one of the leading French astronomers of a century ago (he died in 1923, the year I was born; I knew his second wife, who was herself an expert astronomer). In his book *Popular Astronomy*, published in 1894, Flammarion concluded that Venus was perfectly suited to intelligent life, and went on:

'We may even suppose, without exaggeration, that its inhabitants, organised to live in the midst of these conditions, find themselves at their ease, like a fish in water, and think that our earth is too monotonous and too cold to serve as an abode for active and intelligent beings ... Of what nature are the inhabitants of Venus? Are they endowed with intelligence analogous to our own? Do they resemble us in physical form? Do they pass their life in pleasure, or, rather, are they so tormented by the inclemency of their seasons that they have no delicate perception, and are incapable of any scientific or artistic attention? These are interesting questions, to which we have no reply. All that we can say is, that organised life on Venus must be little different from terrestrial life, and that this world is one of those which resemble ours most. We will not ask then, with the good Father Kircher, whether the water of that world would be good for baptising, and whether the wine would be fit for the sacrifice of Mass; nor, with Huygens, whether the musical

instruments of Venus resemble the harp or the flute; nor, with Swedenborg, whether the young girls walk about without clothing...'

In 1905 Sir Robert Ball, one of Britain's leading astronomers, wrote that "Venus is one of the few globes which might conceivably be the abode of beings not very widely different from ourselves", though he did add that "in view of the composition of the atmosphere there is not much likelihood that Venus is inhabited by any men, women or children resembling those on this Earth".

The idea of advanced life on Venus was taken seriously, at least in some quarters, during the period when Lowell's canal-building Martians were very much in the public eye. C. E. Housden, a strong supporter of Lowell's theories, wrote a book about Venus in 1915, and put forward some views which were, to put it mildly, unusual. He held that the 225-day rotation period is correct, so that convection currents are set up between the day and night hemispheres; deposits of ice and snow are formed just inside the dark hemisphere, and glaciers drift back into the sunlight, enabling the local inhabitants to pump the water back along conduits – these conduits being, of course, the linear streaks shown on Lowell's maps. It is significant that though the august periodical *Nature* carried a review of the book, Housden's theories were merely criticised rather than ridiculed.

Next came Svante Arrhenius, a Swedish physicist whose work was good enough to earn him a Nobel Prize. In his book *Worlds in the Making*, published in 1918, he gave a vivid and highly attractive picture of Venus, which he believed to be a world similar to the Earth of over 200 million years ago – in the Carboniferous Period, when the Coal Forests were being laid down and the most advanced life-forms were amphibians; even the great dinosaurs lay in the future. Arrhenius' description of Venus is worth quoting at some length:

Improbable but charming life forms of Venus, as imagined by Gertrude Moore.

'The average temperature there is calculated to be about 47°C ...
The humidity is probably about six times the average of that on the
Earth, or three times that in the Congo, where the average
temperature is 26°C. The atmosphere of Venus holds about as
much water vapour 5 kilometres above the surface as does the
atmosphere of the Earth at the surface. We must therefore conclude
that everything on Venus is dripping wet. The rainstorms, on the
other hand, do not necessarily bring greater precipitation than
with us. The cloud-formation is enormous, and dense rain-clouds
travel as high up as 10 kilometres. The heat from the Sun does not
attack the ground, but the dense clouds, causing a powerful
external circulation of air which carries the vapour to higher

strata, where it condenses into new clouds. Thus, an effective barrier is formed against horizontal air-currents in the great expanses below. At the surface of Venus, therefore, there exists a complete absence of wind both vertically, as the Sun's radiation is blocked by the ever-present clouds above, and horizontally, due to friction. Disintegration takes place with enormous rapidity, probably about eight times as fast as on Earth, and violent rains carry the products speedily downhill, where they fill the valleys and the oceans in front of all river mouths.

'A very great part of the surface of Venus is no doubt covered with swamps, corresponding to those on Earth in which the coal deposits are formed, except that they are about 30°C warmer. No dust is lifted high into the air to lend it a distinct colour; only the dazzling white reflex from the clouds reaches the outside space and gives the planet its remarkable, brilliantly white lustre. The powerful air-currents in the highest strata of the atmosphere equalise the temperature difference between poles and equator almost completely, so that a uniform climate exists all over the planet analogous to conditions on the Earth during its hottest periods.

'The temperature on Venus is not so high as to prevent luxuriant vegetation. The constantly uniform climatic conditions which exist everywhere result in an entire absence of adaptation to changing exterior conditions. Only low forms of life are therefore represented, mostly no doubt belonging to the vegetable kingdom; and the organisms are of nearly the same kind all over the planet. The vegetative processes are greatly accelerated by the high temperature. Therefore, the lifetime of organisms is probably short. Their dead bodies, decaying rapidly if lying in the open air, will fill it with stifling gases; if embedded in the slime carried down by the rivers, they speedily turn into small lumps of coal, which later, under the pressure of new layers combined with high temperature become particles of graphite ... The temperature at the poles of Venus is probably somewhat lower, perhaps by about 10°C, than the average temperature on the planet. The organisms there should have developed into higher forms than elsewhere, and progress and

culture, if we may so express it, will gradually spread from the poles toward the equator. Later, the temperature will sink, the dense clouds and gloom disperse, and some time, perhaps not before life on Earth has reverted to its simpler forms or even become extinct, a flora and fauna will appear, similar in kind to those which now delight our human eye, and Venus will then indeed be the "Heavenly Queen" of Babylonian fame, not because of her radiant lustre alone, but as the dwelling-place of the highest beings in our Solar System.'

Arrhenius was not alone in his views, but at that time, of course, nobody had any real idea of the nature of the surface. Oceans did not seem improbable, so why not vegetation? One supporter of this idea was G. A. Tikhov, a well-known astronomer in what was then the USSR. In 1955 he suggested that the yellowish colour of Venus could be due to vast tracts of plant life. Because of the high temperature, the planet "must reflect all the Sun's heat rays, of which those visible to the eye are from red to green inclusive. This gives the plants a yellow hue. In addition, the plants must radiate red rays, with the yellow; this gives the planets an orange colour … Thus we get the following gamut of colour: on Mars, where the climate is rigorous, the plants are blue. On Earth, where the climate is intermediate, the plants are green, and on Venus, where the climate is hot, the plants have an orange colour." Tikhov supported the idea that Venus today is rather like the Earth may have been a hundred million or more years ago.

In the same year – 1955 – Sir Fred Hoyle proposed an entirely different theory. Instead of Arrhenius' wet world, the orange vegetation of Tikhov or the increasingly popular dust-bowl picture, he proposed that there might be oceans of oil! He reasoned as follows:

'Suppose an enormous quantity of oil were to gush to the Earth's surface; what would the effect be? The oil, consisting as it does of hydrocarbons, would proceed to absorb oxygen from the air. If the amount of oil were great enough, all the oxygen would be

removed. When this happened the water vapour in our atmosphere would no longer be protected from the disruptive effect of ultraviolet light from the Sun. So water vapour would begin to be disassociated into separate atoms of oxygen and hydrogen. The oxygen would combine with more oil, while the hydrogen atoms would proceed to escape altogether from the Earth out into space. More and more of the water would be dissociated and more and more of the oil would become oxidised. The process would only come to an end when either the water or the oil became exhausted. On the Earth it is clear that water has been dominant over oil. On Venus the situation seems to have been the other way round; the water has become exhausted and presumably the excess of oil remains – just as the excess of water remains on the Earth.

'This possibility has an interesting consequence. In writing previously about the clouds I said that the only suggestion that seemed to fit the observations was that the clouds are made up of fine dust particles. To this suggestion we must now add the possibility that the clouds might consist of drops of oil – that Venus may be draped in a kind of perpetual smog.'

And with regard to the slow axial rotation:

'It is thus reasonable to suppose that the slowing-down of Venus can be explained by the friction of tides – if Venus possesses oceans, but not I think otherwise. Previously, the difficulty was to understand what the liquid oceans were made of. Now we see that the oceans may well be oceans of oil. Venus is probably endowed beyond the dreams of the richest Texas oil-king.'

In view of the present world situation, this would make Venus an attractive target! But at about the same time (1955), two eminent American astronomers, F. L. Whipple and D. H. Menzel, came out with a picture which was different again. This time the Venus oceans were nothing more nor less than ordinary water, with clouds made up of H_2O.

This theory was based essentially on a series of measurements of the polarisation of the light of Venus, made in 1929 by Bernard Lyot in France. Among the substances with which he was able to experiment, Lyot found that only water droplets agreed reasonably well with the variation of polarisation with scattering angle on Venus. In their original paper, Whipple and Menzel wrote:

> *'Lyot's polarization measures indicate that water droplets fit his data satisfactorily. The droplets of the Venusian clouds are uniform in dimension and large enough for high-level airborne dust. No one has been able to to suggest, in place of water droplets, a likely substitute material that would both be available and agree with the polarization and other reflection characteristics observed on the clouds.'*

They therefore suggested that Venus was likely to be completely covered with water. There was a somewhat bizarre corollary. If there were oceans, they would have been formed by the atmospheric carbon dioxide producing seas of soda water.

Actually, the idea of Venus oceans was not new. As we have seen, efforts had been made to explain the Ashen Light by phosphorescent water; in 1924 Sir Harold Jeffreys had written that "Venus may have an ocean with shallow seas, much like ours". On the other hand, H. C. Urey maintained that in all probability Venus once had extensive oceans, but by now all the water had been lost, leaving Venus a sterile world:

> *'The presence of carbon dioxide in the planet's atmosphere is very difficult to understand unless water were originally present, and it would be impossible to understand if water were present now.'*

Well – we have proved that there are no seas, either of water or of oil. The flight of Mariner 2 was enough to show that. So, taking all the evidence into account, can we expect to find any life on Venus?

The instinctive answer is "No", but in view of recent discoveries we must be cautious, because it has been found that living creatures can survive in the most unexpected places. In particular there are the hydrothermal vents, more commonly known as black smokers, which lie in the depths of our oceans. They were first identified as recently as 1977 near the Galapagos Islands, by a scientific team on the research vessel *Alvin*. On the ocean floor, water which has been fiercely heated by the underlying magma flows out through a crack along a rift or a ridge; it really is scalding, with a temperature which may be as high as 350°C (660°F). Sulphur-rich minerals crystallise out on to the rocks, and as they meet the cold water they produce dark, billowing clouds (hence the nickname of "black smokers"); large, chimney-

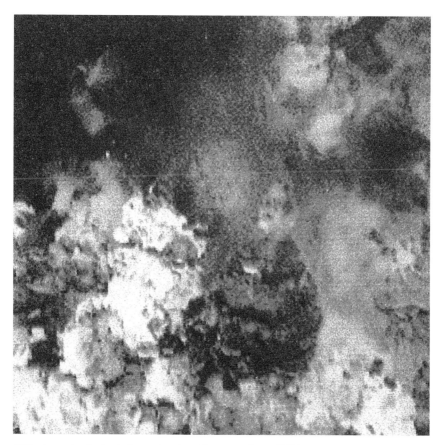

"Black smoker" a hydrothermal vent on the ocean floor off Okinawa.

The early Venus with oceans that were heated and boiled away as the sun became more powerful. Impression by Paul Doherty.

like structures are built up. No environment could be less promising for life – and yet life exists in abundance. There are unique crabs, giant tube-worms, shrimps, yellow mussels, sea-urchins, clams – all in a region where there is no sunlight or available free oxygen, and where the pressure is tremendous. Familiar life-forms could not survive there for a moment, and yet the hydrothermal vents are positively teeming with living creatures.

Venus may have no oceans, but the conditions there are no more hostile than those in the black smokers, and this is why we must not be too quick to rule out all life. On the other hand, if it exists at all it must be very lowly. Science fiction writers delight in populating other worlds with entirely alien beings, but all the available evidence indicates that life must depend upon one particular element – carbon – and we cannot expect any living creatures in environments which are totally unsuited to carbon-based life. If this assumption is wrong, then almost all our modern science is wrong too, which I for one refuse to believe.

Is it possible that life did appear on Venus in the early days, when the Sun was much less powerful than it is now – and if so, could any remnant of it survive? It is not out of the question, but the evidence does seem to be very much against it. More probably Venus has always been sterile. In the future we will certainly obtain specimens for analysis, but if we find anything even remotely resembling a fossil most people – including me – will be very surprised indeed.

None of this can daunt the people whom I always call Independent Thinkers, who are always ready to depart from convention and are supremely confident of their facts. So before summing up, let us digress briefly to enter the mad, mad world of astrology, flying saucers and alien visitors from Outer Space.

14

En Passant

Flying Saucers have been with us now for more than fifty years, and "sightings" have been numerous and varied. In fact, Venus has been responsible for a high percentage of them, including one by President Carter of the United States. There have also been reports of alien landings, and claims that people have actually been kidnapped by "Venusians" (or Martians, or Saturnians). Conspiracy theories are rife, and the Saucerers – or "Ufologists", as they now prefer to be called – are convinced that world governments are doing their best to conceal the fact that we are regularly visited.* It is all great fun, and, in most cases (not all) completely harmless.

Flying saucers made their entry in 1947, when an aircraft pilot named Kenneth Arnold, flying over the United States, reported nine round objects moving at high speed, and passing within a few miles of him. He said that they were flat, like pie-pans, and the term "flying saucer" came from this.

Could they have been alien spacecraft? The idea took hold of popular imagination, and it was inevitable that before long there would be accounts of actual meetings with our cosmic friends. The first, and best known, was due to George Adamski and his co-author Desmond Leslie, who produced a best-selling book entitled *Flying Saucers Have Landed*. George ran a café on the slopes of Palomar Mountain, site of what was then the world's largest

---

* There are even people who believe that nobody has ever landed on the Moon, and that the entire Apollo programme is an elaborate hoax by NASA. All one can really say about these curious folk is that if ignorance is bliss, they must be very happy!

telescope (the Hale 200-inch reflector), but the encounter took place on the Californian desert, near Parker in Arizona.

Together with some friends, George saw the Saucer on terra firma; alone, he located the pilot, who was very like an Earthman apart from the fact that his clothing was rather atypical for a desert expedition, and he was wearing what seemed to be ski trousers; his shoulder-length hair was beautifully waved. Unfortunately he could not talk English (or even American), so that George had to depend upon semaphore, but he was soon able to establish that the Saucer came from Venus, and the two had a very amicable chat. At the end of the interview the Saucer itself came along to join the party, and hovered above the ground. Inside it were several other people. It was in this remarkable craft that the visitor made his final exit, leaving George alone in the desert.

George Adamski became world-famous almost overnight, and remained so until his death more than ten years later. Three years after that initial encounter he wrote a second book, in which he described how he met three Saucer pilots in the lobby of an hotel and was taken for a ride – in the literal, not the metaphorical, sense. It was during this flight that he by-passed the Moon, and saw interesting creatures prancing round on the far hemisphere which is always turned away from the Earth. One member of the crew – a Martian, this time – even explained how the Saucers worked. The written account given by George was detailed, and included a photograph which reminds one of a lampshade with three bulbs underneath (suggestions that it *was* a lampshade with three bulbs underneath I dismiss as unworthy). According to the Martian, these are hollow balls which collect and condense the static electricity sent to them from the magnetic pole. This power is, of course, present all over the universe; one of its natural manifestations is known to us as lightning.

Altogether George went on several flips, both in this and in a larger, more ambitious craft built upon the planet Saturn. By then his hosts had decided to speak English rather than waste time in semaphore, and all sorts of illuminating facts emerged. Eventually the interplanetary visitors departed for home, though not before

George had been the guest of honour at a farewell banquet held in the Saturnian Saucer.

Countless other Saucer books soon appeared. In one of these, the author described his adventures on a Saucer piloted by a beautiful lady from Venus named Mrs Aura Raynes. This was carefully entitled "Non-fiction; an account of a true experience". Sadly, the author's wife sued him for divorce, cited the lady from Venus – and won!

I cannot resist introducing the Aetherius Society, founded in London by a Mr George King, who was then a taxi-driver but has since transferred to Los Angeles as the Rev. Dr George King, official delegate of the Interplanetary Parliament which meets on Saturn, the most highly civilized of the planets. According to a report in the Society's periodical, *Cosmic Voice*, there was a serious crisis in the 1950s, when unfriendly fish-men from the other side of the Galaxy launched a missile at us with the set intention of removing us from the face of the universe. This was because the fish-men lived on a planet which was drying up, and they considered it wholly justifiable to land on Earth and take possession of our oceans. Fortunately the news filtered through to the Interplanetary Parliament, and it was decided – we hope unanimously – to take action. The Master Aetherius, who lives on Venus and has special responsibility for terrestrial affairs, was away at the time, but the Martians were on the alert. With commendable efficiency they hurled a thunderbolt at the missile, and destroyed it. It is not now thought that there will be another attack in the foreseeable future, but in any case we can depend upon the Aetherius Society to give us due warning.

Venus is an extremely advanced planet, with a magnificent Crystal Temple which radiates immense power during a Spiritual Push. According to the descriptions given in *Cosmic Voice*, a typical inhabitant of Venus is from seven to nine feet tall, with long white or fair hair, a cinnamon-coloured skin, and eyes without pupils. Their feet seem abnormally small to carry so tall a body, but of course the gravity on Venus is rather weaker than that of Earth. Aetherius may well look like this, but no specific description of him has been given; we must simply wait until he decides to visit us in person, as no doubt he will do in the foreseeable future.

Some of the Venus-built Saucers are large, and a "mother-ship", capable of carrying up to 7,000 Scout ships, may be as much as 5,000 miles in length. Each is propelled by a magnetic device which exerts an equal thrust on every atom of substance, thereby cancelling out the effects of gravitation. Apparently all this is perfectly simple when one really puts one's mind to it.

Finally, would you like to buy a plot of land on Venus? If so, all you have to do is to e-mail the Lunar Embassy, based in Michigan, USA. For a trifling sum they will provide you with a Venusian Bill of Rights, a site map, and a certificate of ownership. So if you feel like acquiring a "des res" on the slopes of Mount Xoxhiquetzal, there should be no problem...

Life would be much less entertaining without the Independent Thinkers. But whether we will ever be able to go to Venus is another matter, and I fear that we have no chance of being met by a Venusian welcoming committee. The climate is a little too hot.

And after these asides – back to science!

15

Venus in the Future

The Space Age began less than fifty years ago. Since the epic flight of Sputnik 1 we have sent unmanned probes to all the planets except Pluto, we have made controlled landings on Mars, Venus and the little asteroid Eros; men have walked on the Moon, and set up temporary bases there, and if all goes well there is no reasonable doubt that Mars will be reached within the next half-century. But Venus is a very different proposition, and the problems of manned flight there are Himalayan. How do you deal with the intolerable heat, the crushing pressure and the acid clouds? Moreover, can the risks of a manned expedition be reduced to an acceptable level?

It is safe to say that no attempt will be made yet awhile. The next step must be to obtain samples of the surface material, and find out whether there is, or has been, any trace of life there. (I would say that the odds are at least 10,000 to 1 against, but I may be wrong.) Note also that all samples brought back from Venus will have to be kept in strict isolation until we can be sure that they contain nothing harmful, and very prudently even the first two Apollo missions were treated with great caution; all the returning astronauts were quarantined until exhaustive tests had been made. Quarantining was abandoned after Apollo 12 only because it had become absolutely clear that there was no danger. The chances of bringing back harmful contamination from Venus are very slight, but they are not nil. Remember Professor Quatermass!

There have been many suggestions that we might be able to alter the conditions on Venus and change it into a more friendly world, a process known as "terraforming". This may well be tried with Mars, but with Venus it is much more difficult. It is true that Venus' escape velocity is much the same as ours, so that it is capable of holding down an Earth-type atmosphere despite the high temperature, and it is also true that there is plenty of oxygen; the problem is that the oxygen is locked up in the molecules of carbon dioxide and sulphuric acid. It seems that the only hope is to "seed" the clouds, releasing the oxygen; this would also reduce the greenhouse effect, and lower the surface temperature to an acceptable level. But with our present-day technology all this seems very futuristic, and quite beyond our capability. Whether it will always remain so we cannot tell; making long-range forecasts is always unwise.

Venus, like the Earth, has a limited life-span. As we have seen, the Sun is shining because of the nuclear reactions going on deep inside it; hydrogen is being converted to helium, with release of energy and loss of mass. Eventually the supply of available hydrogen will be exhausted, and the Sun will change its structure. In a few thousand million years it will swell out to become a red giant star, as Arcturus is now, and for a while it will increase its luminosity by a factor of at least one hundred, with fatal results for the inner planets. Mercury and Venus are bound to be destroyed; even if the Earth survives, it will be as a molten, uninhabitable mass. It all sounds rather depressing, but at least nothing is likely to happen for a very long time yet.

The first man on Mars has almost certainly been born, though when he will emulate Neil Armstrong and take "one small step" on the Martian surface remains to be seen. We must wait much longer for the first journey to Venus, but there is no harm in speculating, so let me end by inviting you to join me on a trip there. What will we experience and what will we see?

The top of Venus' atmosphere lies about 250 miles above ground level. As we approach in our spacecraft, the clouds below appear as a dense haze, totally concealing the great volcanoes, craters, plateaux and canyons that we know exist. Even from the top of the main atmosphere, at 170 miles, the view is much the same. The temperature here is a modest -27°C (-17°F); at 60 miles up it drops to -90°C (-130°F), but then begins to rise again as the atmosphere becomes denser. At 43 miles up we enter the first clouds, and visibility begins to decrease. By 40 miles the Sun is perceptibly dimmed; by the time we have dropped to 40 miles above ground level the Sun itself is no longer a sharp disk, and is nothing more than a brilliant glare behind the diffuse, yellow cloud-layer made up of tiny particles of sulphuric acid. By now the visibility has fallen to below 4 miles; the temperature is 13°C (55°F), and the outside pressure is about half that of the Earth's air at sea-level.

As we descend, the visibility becomes less, and the temperature rises at an alarming rate. At 30 miles above ground the visibility is only about $1\frac{1}{2}$ miles, and the temperature is 20°C (68°F). We pass quickly through a cloud layer, and then enter the lower regions; the pressure has risen to 1 atmosphere, and the temperature has soared to 200°C (390°F). This is in fact the region where the clouds are densest, and only here would they look at all like the cloud structures on Earth. Below this layer we come to a second clear space, with nothing more than what seems to be haze. Even this clears away at 20 miles above ground, and the visibility has increased to 50 miles; the illumination is about the same as that of a bright, cloudy day on Earth, but the outside temperature is now over 300°C (570°F). We drop down to 12 miles above ground, and visibility is reduced to 12 miles, with a fearsome outside temperature of 380°C (720°F). The outside light is reddish, and we become acutely conscious of the inferno-like conditions we are about to meet. From 4 miles above ground some surface features start to become visible in the red glow – and finally we land; the temperature is now around 470°C, or over 900°F, and the pressure is a crushing 91 atmospheres. Visibility is down to less than two miles; the Sun is not to be seen, and its location can be traced only by an ill-defined, baleful glare. To step outside our spacecraft would be to invite disaster...

Will this ever happen? Quite possibly – but not in our time. At least we can admire Venus as it shines down from the evening twilight or the morning dawn, and it is easy to see why our ancestors named it in honour of the goddess of love and beauty.

Appendix 1
Numerical Data

Distance from the Sun, miles:	max 67,700,000
	mean 67,246,000
	min 66,749,000
Sidereal period:	224.701 days
Synodic period:	583.9 days
Rotation period:	243.018 days
Mean orbital velocity:	22 miles per second
Axial inclination:	178°
Orbital eccentricity:	0.0167
Orbital inclination:	3°23'39".8
Diameter:	7,523 miles
Oblateness:	negligible
Apparent diameter from Earth:	max 65".2, mean 37".3, min 9".5.
Maximum magnitude:	-4.4
Mass, Earth=1:	0.815
Volume, Earth=1:	0.86
Density, water=1:	5.25
Escape velocity:	6.44 miles per second
Surface gravity, Earth=1:	0.909
Albedo:	0.76
Mean surface temperature: cloud-tops	-35°C
surface	467°C
Mean diameter of Sun, as seen from Venus:	44'15"
Solar day:	117 Earth-days (i.e. a daylight period of 58.5 days, followed by a night of equal length.)

Appendix 2
Phenomena of Venus, 2001–2007

E. elongation	Inferior conjunction	W. elongation	Superior conjunction
2001 Jan 17	2001 Mar 29	2001 June 8	(2000 June 11)
2002 Aug 22	2002 Nov 1	2003 Jan 11	2002 Jan 14
2004 Mar 9	2004 June 8	2004 Aug 17	2003 Aug 18
2005 Nov 3	2006 Jan 13	2005 Mar 25	2005 Mar 31
2007 June 9	2007 Aug 18	2007 Oct 28	2006 Oct 27

Appendix 3
Some Estimated Rotation Periods

Year	Observer	Method	Period (x-synchronous)
1667	G. D. Cassini	Visual	23h 21m
1727	F. Bianchini	Visual	24d 8h
1740	J. J. Cassini	Visual	23h 20m
1789	J. Schröter	Visual	23h21m19s
1811	J. Schröter	Visual	23h 21m 7s.977
1881	W. F. Denning	Visual	23h 21m
1880	G. V. Schiaparelli	Visual	*
1892	E. Trouvelot	Visual	24h
1894	C. Flammarion	Visual	24h
1897	H. McEwen	Visual	23h 30m
1900	A. Belopolski	Spectrographic	24h 42m
1903	V. M. Slipher	Spectrographic	*
1909	P. Lowell	Visual and spectrographic	* *
1916	C. Housden	Visual	*
1921	W. H. Pickering	Visual	2d 10h
1924	W. H. Steavenson	Visual	8d
1927	F. E. Ross	Photographic	30h
1932	R. Barker	Visual	*
1934	E. M. Antoniadi	Visual	Very long, or *
1949	V. V. Volkov	Visual	2d 12h
1953	G. D. Roth	Visual	15h
1954	G. P. Kuiper	Photographic	A few weeks
1955	A. Dollfus	Visual and photographic	*
1956	J. D. Kraus	Radio	22h 17m
1962	R. L. Carpenter	Radar	About 250d, retrograde
1964	I. I. Shapiro	Radar	247d, retrograde
1979	I. I. Shapiro	Radar	243d.01, retrograde
	Modern value:	Radar	243d.018, retrograde

Appendix 4
Spacecraft to Venus, 1961–2001

Name	Launch date	Encounter date	Closest approach (miles)	Capsule landing area Lat., Long.
Venera 1	12 Feb 1961	19 May 1961	60,000	—
Mariner 1	22 July 1962	—	—	—
Mariner 2	27 Aug 1962	14 Dec 1962	21,650	—
Zond 1	2 April 1964	?	—	— —
Venera 2	12 Nov 1965	27 Feb 1966	14,900	—
Venera 3	16 Nov 1965	1 Mar 1966	Landed	?
Venera 4	12 June 1967	18 Oct 1967	Landed	+19, 038 (Eistla Regio)
Mariner 5	14 June 1967	19 Oct 1967	—	—
Venera 5	5 Jan 1969	16 May 1969	Landed	-03, 018 (E of Navka Plan.)
Venera 6	10 Jan 1969	17 May 1969	Landed	-05, 023 (E of Navka Plan.)
Venera 7	17 Aug 1970	15 Dec 1970	Landed	-05, 351 (E of Navka Plan.)
Venera 8	26 Mar 1972	22 July 1972	Landed	-10, 355 (E of Navka Plan.)
Mariner 10	3 Nov 1973	5 Feb 1974	5,800	—
Venera 9	8 June 1975	21 Oct 1975	Landed	+31.7, 290.8 (Beta Regio)
Venera 10	14 June 1975	25 Oct 1975	Landed	+16, 291 (Beta Regio)
Pioneer Venus 1	20 May 1978	4 Dec 1978	90	—
Pioneer Venus 2	8 Aug 1978	4 Dec 1978	Landed (9 Dec)	
		Large probe		+04.4, 304.0 (Beta Regio)
		North probe		+59.3, 004.8 (Ishtar Regio)
		Day probe		+31.7, 317.0 (Themis Regio)
		Night probe		-28.7, 056.7 (N of Aino Plan.)
		Bus		-37.9, 290.9 (Themis Regio)
Venera 11	9 Sept 1978	25 Dec 1978	Landed	-14, 299 (Navka Plan.)
Venera 12	14 Sept 1978	22 Dec 1978	Landed	-07, 303.5 (Navka Plan.)
Venera 13	30 Oct 1981	1 Mar 1982	Landed	-07.6, 308 (Navka Plan.)
Venera 14	4 Nov 1981	5 Mar 1982	Landed	-13.2, 310.1 (Navka Plan.)
Venera 15	2 June 1983	10 Oct 1983	620	—
Venera 16	7 June 1983	16 Oct 1983	620	—

Results

Contact lost at 4,700,000 miles from Earth
Total failure; fell in sea
Fly-by. Contact lost on 4 Jan 1963.
Contact lost in a few weeks
In solar orbit.
Lander crushed during descent
Data transmitted during descent
Fly-by. Data transmitted
Lander crushed during descent
Lander crushed during descent
Transmitted for 23 min after landing
Transmitted for 50 min after landing
Data transmitted. En route to Mercury
Transmitted for 55 min after landing – 1 picture
Transmitted for 65 min after landing – 1 picture
Orbiter, 90-41,000 miles, period 24h.
Contact lost 9 Oct 1992
Multiprobe. "Bus" and 4 landers

No transmission after landing
No transmission after landing
Transmitted for 67 min after landing
No transmission after landing
Crash-landing
Transmitted for 95 min after landing
Transmitted for 60 min after landing
Transmitted for 60 min. Soil analysis
Transmitted for 60 min. Soil analysis
Polar orbiter, 620-41,000 miles. Radar mapper
Polar orbiter, 620-41,000 miles. Radar mapper

Appendix 4 (cont'd)
Spacecraft to Venus, 1961–2001

Name	Launch date	Encounter date	Closest approach (miles)	Capsule landing area Lat., Long.
Vega 1	15 Dec 1984	11 June 1985		—
			Lander	+08.5, 176.9 (Rusalka Plan.)
Vega 2	20 Dec 1984	15 June 1985		—
			Lander	-07.5, 179.8 (Rusalka Plan.)
Magellan	5 May 1989	10 Aug 1990	183	—
Galileo	18 Oct 1989	10 Feb 1990	9940	—
Cassini	18 Oct 1997	26 Apr 1998	177	—
		20 Jun 1999	372	

Results

Fly-by; en route to Halley's Comet
Lander transmitted for 20 min after arrival.
Balloon dropped into Venus' atmosphere
Fly-by; en route to Halley's Comet
Lander transmitted for 21 min after arrival.
Balloon droppedinto Venus' atmosphere
Orbiter. 183-5,250 miles. Radar mapper. Burned away in
Venus' atmosphere 11 Oct 1994
Fly-by; en route to Jupiter
Two fly-bys, en route to Saturn

Appendix 5
Features on Venus

Many hundreds of features on Venus have now been named. Of course they cannot be seen from Earth, and a list of just a few of them is given here for the sake of completeness.

CRATERS

Name	Lat.	Long. E	Diameter, (miles)
Addams	56.1 S	98.0	53
Baker	62.6 N	40.5	62
Cleopatra	65.9 N	07.0	65
Cochran	51.8 N	143.2	62
Dickinson	74.3 N	177.3	43
Erxleben	59.9 S	039.4	19
Isabella	29.8 S	204.2	103
Joliot Curie	16.8 S	062.5	62
Marie Celeste	28.5 N	140.2	59
Mead	12.5 N	057.2	174
Meitner	55.6 S	321.6	93
Mona Lisa	25.5 N	025.1	53
Warren	11.8 S	176.5	31
Yablochkina	48.2 N	195.1	39

CHASMATA

Name	Lat.	Long. E	Length (miles)
Artemis	41.2 S	138.5	1,919
Dali	17.6 S	167.0	1,290
Devana	09.6 N	284.4	1,004
Heng-O	06.6 N	355.5	456
Ix Chel	10.0 S	073.4	313
Parga	24.5 S	271.5	1,162

CORONAE

Name	Lat.	Long. E	Diameter (miles)
Artemis	35.0 S	135.0	1,616
Ceres	16.0 S	151.5	420
Heng-O	02.0 N	355.0	659
Quetzalpetlatl	64.0 S	354.4	249

DORSA

Name	Lat.	Long. E	
Absonnutli	47.9 N	194.8	1,061
Frigg	51.2 N	148.9	557
Iris	52.7 N	221.3	1,274
Nephele	39.6 N	139.0	1,204
Tezan	81.4 N	047.1	671
Saule	58.0 S	206.0	855

FOSSAE

Name	Lat.	Long. E	Diameter, (miles)
Arionrod	37.0 S	239.9	444
Hildr	45.4 N	159.4	421
Nike	63.0 S	347.0	528

MONS

Name	Lat.	Long. E	Diameter, (miles)
Gula	21.9 N	359.1	172
Maat	00.5 N	194.6	245
Ozza	04.5 N	201.0	315
Rhea	32.4 N	282.2	135
Sapas	08.5 N	188.3	135
Sekmet	44.2 N	240.8	210
Sif	22.0 N	352.4	124
Theìa	22.7 N	281.0	140
Ushas	24.3 S	324.6	253
Xochiquetzal	03.5 N	270.0	50

MONTES

Name	Lat.	Long. E	Diameter, (miles)
Akna	68.9 N	318.2	516
Danu	57.5 N	34.0	502
Freyja	74.1 N	333.8	360
Maxwell	65.2 N	003.3	495

PATERA

Name	Lat.	Long. E	Diameter, (miles)
Boadicea	56.0 N	096.0	137
Broswitha	35.8 N	034.8	101
Raskova	51.0 S	222.8	50
Razia	46.2 N	197.8	98
Sacajawea	64.3 N	335.4	145
Sappho	14.1 N	016.5	57

PLANITIA

Name	Lat.	Long. E	Diameter, (miles)
Aino	40.5 S	094.5	3,097
Attalanta	45.6 N	165.8	1,273
Guinevere	21.9 N	325.0	4,673
Helen	51.7 S	263.9	2,711
Lavinia	47.3 S	347.5	1,753
Leda	44.0 N	065.1	1,796
Navka	08.1 S	317.6	1,311
Niobe	21.0 N	112.3	3,112
Rusaika	09.8 N	170.1	2,272
Sedna	42.7 N	340.7	2,220

PLANUM

Name	Lat.	Long. E	Diameter, (miles)
Lakshmi	68.6 N	339.3	1,456

REGIONES

Name	Lat.	Long. E	Diameter, (miles)
Alpha	25.5 S	001.3	1,179
Asteria	21.6 N	267.5	703
Beta	25.3 N	282.8	1,783
Eistla	10.5 N	021.5	4,981
Metis	72.0 N	256.0	453
Ovda	02.8 S	085.6	3,282
Phoebe	06.0 S	282.8	1,773
Tethus	66.0 N	120.0	1,498
Themis	37.4 S	284.2	1,126
Thetis	11.4 S	129.9	1,741
Ulfrun	20.5 N	223.0	2,457

RUPES

Name	Lat.	Long. E	Diameter, (miles)
Fornax	30.3 N	201.1	453
Uosar	76.8 N	341.2	516
Ut	55.3 N	321.9	420
Vesta	58.3 N	323.9	490

TERRAE

Name	Lat.	Long. E	Diameter, (miles)
Aphrodite	05.8 N	104.8	6,214
Ishtar	70.4 N	027.5	3,486
Lada	54.4 S	342.5	5,354

TESSERAE

Name	Lat.	Long. E	Diameter, (miles)
Ananke	53.3 N	133.3	659
Dekia	57.4 N	071.8	847
Fortuna	69.9 N	045.3	1,740
Laima	55.0 N	048.5	603
Manzan-Gurme	39.5 N	359.5	841
Meshkenet	65.8 N	103.1	657
Tellus	42.6 N	076.8	1,447

THOLUS

Name	Lat.	Long. E	Diameter, (miles)
Ashtart	48.7 N	247.0	86
Semele	64.3 N	202.0	121
Upunusa	66.2 N	242.4	139

UNDAE

Name	Lat.	Long. E	Diameter, (miles)
Al-Uzza	67.7 N	090.5	93
Menat	24.8 S	339.4	16
Nirgal	09.0 N	060.7	140

VALLIS

Name	Lat.	Long. E	Diameter, (miles)
Baltis	37.3 N	161.4	3,730
Bennu	01.3 N	341.2	491
Citialpui	57.4 S	185.0	1,460
Kallistos	51.1 S	021.5	559
Saga	76.1 N	340.6	280
Vakarine	05.0 N	336.4	388
Ymoja	71.6 S	204.8	242

Bibliography

Cattermole, P. *Venus* UCL Press, 1994

Grady, M. *Search for Life* Natural History Press, 2001

Grinspoon, H. *Venus Revealed* Helix Books, 1997

Hunten, D. L. (ed) *Venus* University of Arizona Press, 1983

Kidger, M. *The Star of Bethlehem* Princeton University Press, 1999

Marov, V., and Grinspoon, H. *The Planet Venus* Yale University Press, 1998

Maunder, M., and Moore, P. *Transit, When Planets Cross the Sun*
  Springer Verlag, 1999

Moore, P. *Astronomy Data Book* Institute of Physics Publishing, 2000

Moore, P. *Atlas of the Universe* Philip, 2000

Moore, P. *The Star of Bethlehem* Canopus Books, 2001

Index